W9-AEN-605

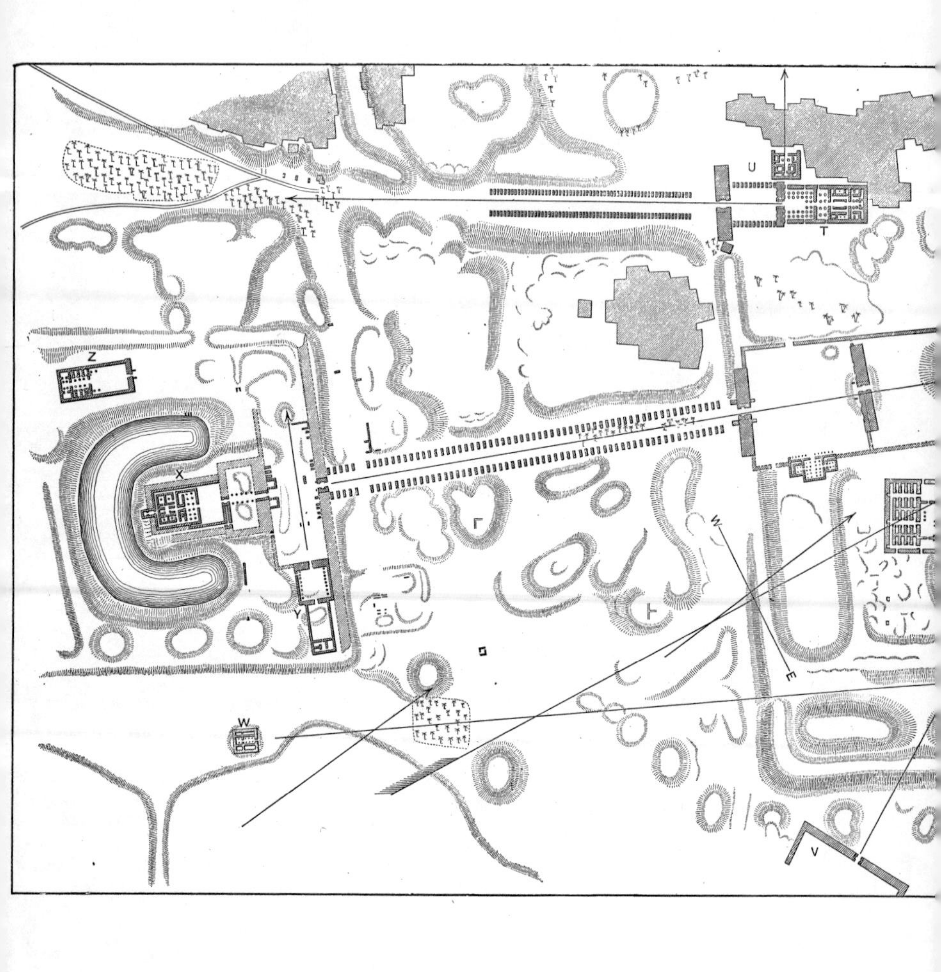

PLAN OF THE TEMPLES AT KARNAK (FROM LEPSIUS), SHOWING THEIR ORIENTATIONS.

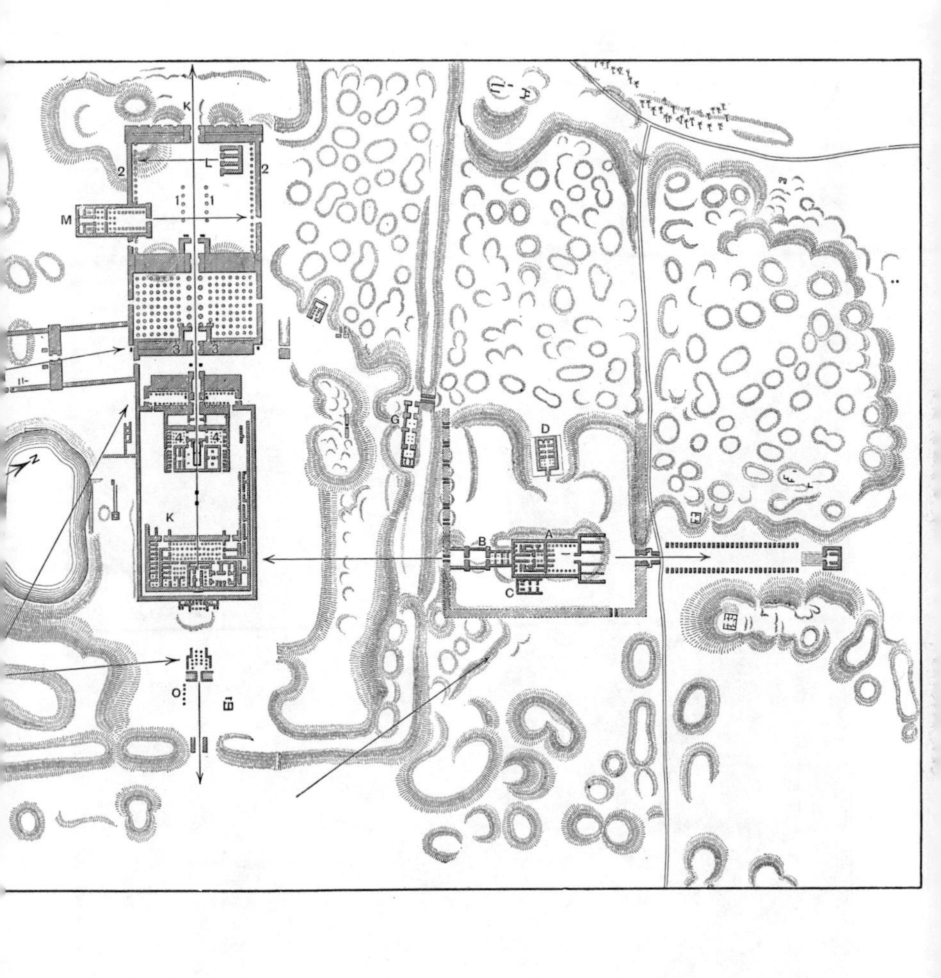
K
L
2
1
M
3
4
K
A
B
C
D
G
O

THE TEMPLE OF AMEN-RĀ, LOOKING FROM THE SANCTUARY TOWARDS THE PLACE OF SUNSET AT THE SUMMER SOLSTICE.

(*From a Photograph by the Author.*)

The Dawn of Astronomy

A Study of the Temple Worship and Mythology of the Ancient Egyptians

by

J. Norman Lockyer

With a Preface by

GIORGIO DE SANTILLANA

The M.I.T. Press

Massachusetts Institute of Technology
Cambridge, Massachusetts

FIRST EDITION 1894
PUBLISHED BY
CASSELL AND COMPANY LIMITED

FIRST MIT PRESS PAPERBACK EDITION, AUGUST 1964
SECOND PRINTING, MARCH 1973

ISBN 0 262 12014 3 (hardcover)
ISBN 0 262 62003 0 (paperback)
LIBRARY OF CONGRESS CATALOG CARD NUMBER 64-22201
PRINTED IN THE UNITED STATES OF AMERICA

PREFACE TO THIS EDITION

Giorgio de Santillana

As this important but almost forgotten book is brought back into print after seventy years, it would be fair first to remind the reader of the author's credentials. Sir Norman Lockyer (1836–1920) was one of the major English astronomers of his time. Born in Rugby, he completed his education on the Continent of Europe, and came to astronomy by way of private study and a clerkship in the War Office. In 1870 he was appointed secretary to the Duke of Devonshire's Royal Commission on science; a few years later, on the foundation of the Royal College of Science in London, he became director of the solar physics observatory and professor of astronomical physics. From 1866, he had been a pioneer in sun and star spectroscopy. He inaugurated *Nature* in 1869 and edited it until his death. His interests went far afield, as we shall see. Always something of a maverick, he suddenly abandoned the Royal College and removed the instruments to his estate at Sidmouth, where he founded his own observatory. It is now the "Lockyer's Observatory" and contains a mass of his papers still unpublished.

Lockyer's fame is solidly based on his study of the sun. In 1868 he described the flares and prominences of the sun as located in a layer he called the chromosphere, and applied the Doppler principle to its movements. In 1868 Lockyer and Janssen, working independently, discovered a spectroscopic method whereby the solar prominences could be studied in daylight, whereas previously they were observable only during a total eclipse. To commemorate this discovery, a medal bearing the names of both astronomers was struck by the French government in 1872. Lockyer received the Rumford medal in 1874, and was vice-president of the Royal Society in 1892–1893. Among his most important discoveries is that of a new element in the solar atmosphere that he called "helium" and that was found later among the rare gases on earth.

After 1890, Lockyer became interested in a problem which had also attracted Newton, that of bringing in astronomy to assist the chronology of history. After a careful investigation of Egyptian monuments on the spot, he published in 1894 his *Dawn of Astronomy*, a work of far-reaching consequences. Egyptologists dismissed it with good-natured laughter, advising the cobbler to stick to his last, and the book dropped out of sight,[1] to the point that when Zaba wrote his important monograph on the orientation of the pyramids[2] in 1953, he was unable to find a copy within reach from Prague and lamented his inability to consult it.

For archeologists, Lockyer was not a member of the guild; despite his fame as an astronomer, they may have taken him for one of the usual excited pyramidologists (Piazzi Smyth, another astronomer, had created a stir a century earlier). There is no doubt that Lockyer showed obviously weak sides, such as his fanciful speculations about the origins of astronomy, or his equally imaginative reconstruction of Egyptian history, ancient totemism, and the like. One is disturbed by his hasty generalizations, as when he ventures to say that Isis stands for anything luminous to the eastward heralding sunrise (p. 293) or that Osiris stands for any celestial body becoming invisible (p. 296); or proposes as an alternative that mummies in hieroglyphs might indicate a setting star, horns and disk a rising one. Such remarks can breed uneasy diffidence, even among his admirers. On the other hand, it is not as arbitrary as it sounds to suspect an equation between Aphrodite, Artemis, and Persephone in cases where no immense architectural graveyard is available to confirm it.

Sir Norman Lockyer's merit lies in the kind of questions he dared to ask, in his unshakable awareness of a veritable technical language hidden in the myths, in his astronomical capacity to decipher it by investigating on the spot. He derived his conclusions about the astronomical character of Egyptian religion on the solidest architectural measurements and astronomical calculations, and then saw how these are represented in the "divine language" of inscriptions and literary texts. From a model like this we can see better how we have to proceed when we have only the latter kind of material to work with, as usually happens.

Lockyer not only had enough confidence to take for certain that the orientation of the temples had astronomical reasons but also used his instruments to work out which of the possible astronomical reasons would stand up under scrutiny.

All of this was beyond the ken of ordinary philologists, and so they chose to ignore it.

Later, Zaba had to start all over again. His main concern was the orientation of the pyramids, which are very accurately to the North and South

[1] Lockyer's later book on Stonehenge (1906), based on admittedly frailer evidence, suffered a similar fate at the hands of Celtic archeologists. A recent professional work (R. F. C. Atkinson, *Stonehenge,* London, 1956) dismissed his astronomical alignments on the summer solstice on plausible archeological grounds. The issue had been considered settled until last year when Professor Gerald Hawkins of Boston University took up the problem again with more alignments and a computer to help him, confirming Lockyer's suggested dates.

[2] Zbynek Zaba, "L'orientation astronomique dans l'ancienne Egypte et la précession de l'axe du monde" (supplement to: *Archiv Orientalni,* Prague, Czechoslovak Academy of Sciences, 1953).

(or East-West). He found it necessary to argue at length against the current opinion that the accuracy is purely coincidental, requiring no precise observational method. To such a pass we have come, and there are by now even historians of science willing to follow. Zaba himself yields to the trend, occasionally, and accepts without examination the opinion that non-meridional temples are oriented by nothing more than practical convenience, "symmetrophobia," and a general preference for the direction of the Nile. Had he been able to read Lockyer, he would have changed his mind.

Lockyer's thought, no doubt, requires deep revision. Yet there is another point that must be made from his work, and it leads to conclusions of the utmost importance. When a stellar temple is oriented so accurately that it requires several reconstructions at intervals of a few centuries, which involve each time the rebuilding of its narrow alignment on a star, and the wrecking of the main symmetry that goes with it; when Zodiacs, like that of Denderah, are deliberately depicted in the appearance they would have had centuries before, as if to date the changes, then it is not reasonable to suppose the Egyptians unaware of the Precession of the Equinoxes, even if their mathematics was unable to predict it numerically. Lockyer lets the facts speak for themselves, but it is he who has given the proof. Actually, the Egyptians do describe the Precession, but in language usually written off as mythological or religious. This is perhaps a habit so deeply ingrained in us after 400 years of the "warfare between religion and science" that we never realize how much it corrupts our judgment when extrapolated into ancient history of other civilizations. And now that by a strange turn of events in our scientific age, the Irrational has won out in the minds of scholars under the fashionable form of the Great Unconscious, the confusion has only become worse confounded.

Lockyer was not too bold, as is usually said; he was not bold enough. Had he lived in the time of Lepsius and Brugsch, he might have found more courage. He would have recognized planetary gods in the documents, had he not been bemused by the current verbiage about cult practices, which is making Egyptian history ever less interesting. The time has come to reopen the case, to honor Lockyer as a pioneer, and to carry on in his spirit, with securer data.

Cambridge, Massachusetts
May, 1964

AUTHOR'S PREFACE

THE enormous advance which has been recently made in our astronomical knowledge, and in our power of investigating the various bodies which people space, is to a very great extent due to the introduction of methods of work and ideas from other branches of science.

Much of the recent progress has been, we may indeed say, entirely dependent upon the introduction of the methods of inquiry to which I refer. While this is generally recognized, it is often forgotten that a knowledge of even elementary astronomy may be of very great assistance to students of other branches of science; in other words, that astronomy is well able to pay her debt. Amongst those branches is obviously that which deals with man's first attempts to grasp the meaning and phenomena of the universe in which he found himself before any scientific methods were available to him; before he had any idea of the origins or the conditionings of the things around him.

In the present volume I propose to give an account of some attempts I have been making in my leisure moments during the past three years to see whether any ideas could be obtained as to the early astronomical views of the Egyptians, from a study of their temples and the mythology connected with the various cults.

How I came to take up this inquiry may be gathered from the following statement: —

It chanced that in March, 1890, during a brief holiday, I went to the Levant. I went with a good friend, who, one day when we were visiting the ruins of the Parthenon, and again when we found ourselves at the temple at Eleusis, lent me his pocket-compass. The curious direction in which the Parthenon was built, and the many changes of direction in the foundations at Eleusis revealed by the French excavations, were so very striking and suggestive that I thought it worth while to note the bearings so as to see whether there was any possible astronomical origin for the direction of the temple and the various changes in direction to which I have referred. What I had in my mind was the familiar statement that in England the eastern windows of churches face generally — if they are properly constructed — to the place of sunrising on the festival of the patron saint; this is why, for instance, the churches of St. John the Baptist face very nearly north-east. This direction towards the sunrising is the origin of the general use of the term *orientation,* which is applied just as frequently to other buildings the direction of which is towards the west or north or south. Now, if this should chance to be merely a survival from ancient times, it became of importance to find out the celestial bodies to which the ancient temples were directed.

When I came home I endeavoured to ascertain whether this subject had been worked out. I am afraid I was a nuisance to many of my archaeological friends, and I made as much inquiry as I could by looking into books. I found, both from my friends and from the books, that this question had not been discussed in relation to ancient temples, scarcely even with regard to churches outside England or Germany.

It struck me that, since nothing was known, an inquiry into the subject — provided an inquiry was possible for a stay-at-home — might help the matter forward to a certain extent. So, as it was well known that the temples in Egypt had been most carefully examined and oriented both by the French in 1798 and by the Prussians in 1844, I determined to see whether it was possible to get any information on the general question from them, as it was extremely likely that such temples as that at Eleusis were more or less connected with Egyptian ideas. I soon found that, although neither the French nor the Germans apparently paid any heed to the possible astronomical ideas of the temple-builders, there was little doubt that astronomical considerations had a great deal to do with the direction towards which these temples faced. In a series of lectures given at the School of Mines in November, 1890, I took the opportunity of pointing out that in this way archaeologists and others might ultimately be enabled to arrive at dates in regard to the foundation of temples, and possibly to advance knowledge in several other directions.

After my lectures were over, I received a very kind letter from one of my audience, pointing out to me that a friend had informed him that Professor Nissen, in Germany, had published some papers on the orientation of ancient temples. I at once ordered them. Before I received them I went to Egypt to make some inquiries on the spot with reference to certain points which it was necessary to investigate, for the reason that when the orientations were observed and recorded, it was not known what use would be made of them, and certain data required for my special inquiry were wanting. In Cairo also I worried my archaeological friends. I was told that the question had not been discussed; that, so far as they knew, the idea was new; and I also gathered a suspicion that they did not think much of it. However, one of them, Brugsch Bey, took much interest in the matter, and was good enough to look up some of the old inscriptions, and one day he told me he had found a very interesting one concerning the foundation of the temple at Edfû. From this inscription it was clear that the idea was not new; it was possibly six thousand years old. Afterwards I went up the river, and made some observations which carried conviction with them and strengthened the idea in my mind that for the orientation not only of Edfû, but of all the larger temples which I examined, there was an astronomical basis. I returned to England at the beginning of March, 1891, and within a few days of landing received Professor Nissen's papers.

I have thought it right to give this personal narrative, because, while it indicates the relation of my work to Professor Nissen's, it enables me to make the acknowledgment that the credit of having first made the suggestion belongs, so far as I know, solely to him.[1]

The determination of the stars to which some of the Egyptian temples, sacred to a known divinity, were directed, opened a way, as I anticipated, to a study of the astronomical basis of parts of the mythology. This inquiry I have carried on to a certain extent, but it requires an Egyptologist to face it, and this I have no pretensions to be. It soon became obvious, even to an outsider like myself, that the mythology was intensely astronomical, and crystallised early ideas suggested by actual observations of the sun, moon, and stars. Next, there were apparently two mythologies, representing two schools of astronomical thought.

Finally, to endeavour to obtain a complete picture, it became necessary to bring together the information to be obtained from all these and other sources, including the old Egyptian calendars, and to compare the early Babylonian results with those which are to be gathered from the Egyptian myths and temple-orientations.

It will, I think, be clear to anyone who reads this volume that its limits and the present state of our knowledge have only allowed me really to make a few suggestions. I have not even attempted to exhaust any one of the small number of subjects which I have brought forward; but if I have succeeded so far as I have gone, it will be abundantly evident that, if these inquiries are worth continuing, a very considerable amount of work has to be done.

Of this future work, the most important, undoubtedly, is a re-survey of the temple sites, with modern instruments and methods. Next, astronomers must produce tables of the rising and setting conditions of the stars for periods far beyond those which have already been considered. The German Astronomical Society has published a table of the places of a great many stars up to 2000 B.C., but to carry on this investigation we must certainly go back to 7000 B.C., and include southern stars. While the astronomer is doing this, the Egyptologist, on his part, must look through the inscriptions with reference to the suggestions which lie on the surface of the inquiry. The astronomical and associated mythological data want bringing together. One part of that work will consist in arranging tables of synonyms like those to which I presently refer in the case of the goddesses. My own impression is

[1]My lectures, given in November, 1890, were printed in *Nature,* April–July, 1891, under the title "On some Points in the Early History of Astronomy," with the following note: — "From shorthand notes of a course of lectures to working men delivered at the Museum of Practical Geology, Jermyn Street, in November, 1890. The notes were revised by me at Aswân during the month of January. I have found, since my return from Egypt in March, that part of the subject-matter of the lectures had been previously discussed by Professor Nissen, who has employed the same materials as myself. To him, therefore, so far as I at present know, belongs the credit of having first made the suggestion that ancient temples were oriented on an astronomical basis. His articles are to be found in the *Rheinisches Museum für Philologie,* 1885."

that this work will not really be so laborious as the statement of it might seem to imply. I have attempted to go over the ground during the last two years as well as my ignorance would allow me, and I have arrived at the impression that the number both of gods and goddesses will be found to be extremely small; that the apparent wealth of the mythology depends upon the totemism of the inhabitants in the Nile valley — by which I mean that each district had its own special animal as the emblem of the tribe dwelling in it, and that every mythological personage had to be connected in some way with these local cults. After this work is done, it will be possible to begin to answer some of the questions which I have only ventured to raise.

I am glad to take this opportunity of expressing my obligations to the authorities in Egypt for the very great help they gave towards the furthering of the inquiries which were set on foot there. Many of my own local observations would, in all probability, never have been made if my friend Major A. Davis, of Syracuse (New York) had not invited me to join him in a cruise up the river in the s.s. *Mohamet Aly*, and practically given me full command of her movements. My best thanks are due to him not only for his hospitality, but for sympathetic aid in my inquiries.

Dr. Wallis Budge and Captain Lyons, R.E., have rendered continual help while this book has been in progress, and I cannot sufficiently thank them; to the first-named I am especially indebted for looking over the proof sheets. I am also under obligations to Professors Maspero, Krall, and Max Müller for information on certain points, and to Professors Sayce and Jensen for many valuable suggestions in the chapters dealing with Babylonian astronomy.

J. NORMAN LOCKYER.

CONTENTS

LIST OF ILLUSTRATIONS.

THE

DAWN OF ASTRONOMY.

CHAPTER I.

THE WORSHIP OF THE SUN AND THE DAWN.

WHEN we inquire among which early peoples we are likely to find the first cultivation of astronomy, whatever the form it may have taken, we learn that it is generally agreed by archæologists that the first civilisations which have so far been traced were those in the Nile Valley and in the adjacent countries in Western Asia.

The information which we possess concerning these countries has been obtained from the remains of their cities, of their temples—even, in the case of Babylonia, of their observatories and of the records of their observations. Of history on papyrus we have relatively little.

Not so early as these, but of an antiquity which is still undefined, are two other civilisations with which we became familiar before the treasure-houses of Egypt and Babylonia were open to our inquiries. These civilisations occupied the regions now called India and China.

The circumstances of these two groups are vastly dissimilar so far as the actual sources of information are concerned; for in relation to China and India we have paper records, but, alas! no monuments of undoubtedly high antiquity. It is true that there are many temples in India in the present day, but, on the authority of Prof. Max Müller, they are relatively modern.

The contrary happens in Egypt, for there monuments exist more ancient than any of the inscribed records; monuments indicating a more or less settled civilisation; a knowledge of astronomy, and temples erected on astronomical principles for the purposes of worship, the astronomers being called "the mystery teachers of Heaven."

We go back in Egypt for a period, as estimated by various authors, of something like 6,000 or 7,000 years. In Babylonia inscribed tablets carry us into the dim past for a period of certainly 5,000 years; but the so-called "omen" tablets indicate that observations of eclipses and other astronomical phenomena had been made for some thousands of years before this period. In China and in India we go back as certainly to more than 4,000 years ago.

When one comes to examine the texts, whether written on paper or papyrus, burnt in brick, or cut on stone, which archæologists have obtained from all these sources, we at once realise that man's earliest observations of the heavenly bodies in all the regions we have named may very fairly be divided into three perfectly distinct stages. I do not mean to say that these stages follow each other exactly, but that at one period one stage was more developed than another, and so on.

For instance, in the first stage, wonder and worship were the prevalent features; in the second, there was the need of applying the observation of celestial phenomena in two directions, one the direction of utility—such as the formation of a calendar and the foundation of years and months; and the other the astrological direction.

Supplied as we moderns are with the results of astronomical observation in the shape of almanacs, pocket-books, and the like, it is always difficult, and for most people quite impossible, to put ourselves in the place and realise the conditions of a race emerging into civilisation, and having to face the needs of the struggle for existence in a community which, in the nature of the case, must have been agricultural. Those would best succeed who best knew when "to plow and sow, and reap and mow;" and the only means of knowledge was at first the observation of the heavenly bodies. It was this, and not the accident of the possession of an extended plain, which drove early man to be astronomically minded.

The worship stage would, of course, continue, and the priests would see to its being properly developed; and the astrological direction of thought, to which I have referred, would gradually be connected with it, probably in the interest of a class neither priestly nor agricultural.

Only more recently—not at all, apparently, in the early stage—were any observations made of any celestial object for the mere purpose of getting knowledge. We know from the recent discoveries of Strassmaier and Epping that this stage was reached at Babylon at least 300 years B.C., at which time regular calculations were made of the future positions of moon and planets, and of such extreme accuracy that they could have been at once utilised for practical purposes. It looks as if rough determinations of star places were made at about the same time in Egypt and Babylonia.

This abstract inquiry is now practically the only source of interest in astronomy to us; we no longer worship the sun; we no longer believe in astrology; we have our calendar; but we must have a Nautical Almanac calculated years beforehand, and some of us like to know a little about the universe which surrounds us.

It is very curious and interesting to know that the first

stage, the stage of worship, is practically missing in the Chinese annals; the very earliest Chinese observations show us the Chinese, a thoroughly practical people, trying to get as much out of the stars as they could for their terrestrial purposes.

In Babylonia it is a very remarkable thing that from the beginning of things—so far as we can judge from the records—the sign for God was a star.

We find the same idea in Egypt: in some of the hieroglyphic texts three stars represented the plural "gods."

I have already remarked that the ideas of the early Indian civilisation, crystallised in their sacred books called Vedas, were known to us long before either the Egyptian or the Babylonian and Assyrian records had been deciphered.

Enough, however, is now known to show that we may take the Vedas to bring before us the remnants of the first ideas which dawned upon the minds of the earliest dwellers in Western Asia—that is, the territory comprised between the Mediterranean, the Black Sea, the Caucasus, the Caspian Sea, the Indus, and the waters which bound the southern coasts—say, as far as Cape Comorin. Of these populations, the Egyptians and Babylonians may be reckoned as the first. According to Lenormant—and he is followed by all the best scholars—this region was invaded in the earliest times by peoples coming from the steppes of Northern Asia. Bit by bit they spread to the west and east. There are strange variants in the ideas of the Chaldæans already recovered from the inscriptions and those preserved in the Vedas. Nevertheless, we find a sun-god[1] and the following hymn:—

"Oh Sun, in the most profound heaven thou shinest. Thou openest the locks which close the high heavens. Thou openest the door of heaven. Oh

[1] Maspero, "Histoire ancienne des Peuples de l'Orient," p. 136.

Sun, towards the surface of the earth thou turnest thy face. Oh Sun, thou spreadest above the surface, like a mantle, the splendour of heaven."

Let us consider for a moment what were the first conditions under which the stars and the sun would be observed. There was no knowledge, but we can very well understand that there was much awe, and fear, and wonder. Man then possessed no instruments, and the eyes and the minds of the early observers were absolutely untrained. Further, night to them seemed almost death—no man could work; for them there was no electric light, to say nothing of candles; so that in the absence of the moon the night reigned like death over every land. There is no necessity for us to go far into this matter by trying to put ourselves into the places of these early peoples; we have only to look at the records: they speak very clearly for themselves.

But the Vedas speak fully, while as yet information on this special point is relatively sparse from the other regions. It is wise, therefore, to begin with India, whence the first complete revelations of this kind came. Max Müller and others during recent years have brought before us an immense amount of most interesting information, of the highest importance for our present subject.

They tell us that 1,500 years B.C. there was a ritual, a set of hymns called the Veda (*Veda* meaning "knowledge"). These hymns were written in Sanskrit, which a few years ago was almost an unknown language; we know now that it turns out to be the nearest relation to our English tongue. The thoughts and feelings expressed in these early hymns contain the first roots and germs of that intellectual growth which connects our own generation with the ancestors of the Aryan races—"those very people who, as we now learn from the Vedas, at the rising and the setting of the sun, listened with trembling hearts to the sacred songs chanted by their priests. The Veda, in fact, is the oldest book in which we can study the first beginnings of our language and of everything which is embodied in all the languages under the sun." The oldest, most primitive, most simple form of Aryan Nature-worship finds expression in this wonderful hymnal, which doubtless brings before us the rituals of the ancient Aryan populations, represented also by the Medes and Persians.

There was, however, another branch, represented by the Zend-Avesta, as opposed to the Vedas, among which there was a more or less conscious opposition to the gods of Nature, to which we are about to refer, and a striving after a more spiritual deity, proclaimed by Zoroaster under the name of Ahura-Mazda, or Ormuzd. The existence of these rituals side by side in time tends to throw back the origin of the Nature-worship of both. Now, what do we find? In the Veda the gods are called Devas, a word which means "bright"; brightness or light being one of the most general attributes shared by the various manifestations of the deity. What were the deities? The sun, the sky, the dawn, fire, and storm. It is clear, in fact, from the Vedas that sunrise was, to those from whom the ritual had been derived, the great revelation of Nature, and in time, in the minds of the poets of the Veda, *deva*, from meaning "bright," gradually came to mean "divine." Sunrise it was that inspired the first prayers of our race, and called forth the first sacrificial flames. Here, for instance, is an extract from one of the Vedas. "Will the sun rise again? Will our old friend the Dawn come back again? Will the power of Darkness be conquered by the God of Light?"

These three questions in one hymn will show what a questionable stage in man's history is thus brought before us, and how the antithesis between night and day was one of the first things to strike mankind. We find very many names for Sun-gods—

> Mitra, Indra (the day brought by the sun),
> Sûrya, Vasishtha, Arusha (bright or red);

and for the Dawn-gods—

> Ushas, Dyaus, Dyotanâ,
> Ahanâ, Urvasī.

We have only to consider how tremendously important must have been the coming of the sun in the morning, bringing everything with it; and the dying away of the sun in the evening, followed at once by semi-tropical quick darkness, to cease to wonder at such worship as this. Here is an extract from one hymn to the Dawn (Ushas):—

"(1) She shines upon us like a young wife, rousing every living being to go to his work; when the fire had to be kindled by men she made the light by striking down darkness.

"(2) She rose up spreading far and wide, and moving everywhere, she grew in brightness, wearing her brilliant garment [the mother of the cows (the mornings)], the leader of the days, she shone gold-coloured, lovely to behold.

"(3) She, the fortunate, who brings the eye of the gods, who leads the white and lovely steed (of the sun), the *Dawn*, was seen revealed by her rays, with brilliant treasures, following everyone.

"(4) Thou art a blessing when thou art near . . . Raise up wealth to the worshipper, thou mighty *Dawn*.

"(5) Shine for us with thy best rays, thou bright Dawn. . . .

"(6) Thou daughter of the sky, thou high-born Dawn. . . ."

In addition to the Sun and the Dawn, which turn out to be the two great deities in the early Indian Pantheon, other gods are to be met with, such as Prithivī, the Earth on which we dwell; Varuna, the Sky; Ap, the Waters; Agni, the Fire; and Maruts, the Storm-gods. Of these, Varuna is especially interesting to us. We read:—

"Varuna stemmed asunder the wide firmament; he lifted up on high the bright and glorious heaven; he stretched out apart the starry sky and the earth."

Again—

"This earth, too, belongs to Varuna, the king, and this wide sky with its ends far apart. The two seas (the sky and the ocean) are Varuna's loins."

Finally, the result of all this astral worship was to give an idea of the connection between the earth and the sun and the heavens, which are illustrated in later Indian pictures, bringing before us modernised and much more concrete views of these early notions, ultimately transformed into this piece of poetic thought, that the earth was a shell supported by elephants (which represent strength), the elephants being supported on a tortoise (which represents infinite slowness).

This poetical view subsequently gave way to one less poetical—namely, that the earth was supported by pillars; on what the pillars rested is not stated, and it does not matter. We must not consider this as ridiculous, and pardonable merely because it is so early in point of time; because, coming to the time of Greek civilisation, Anaximander told us that the earth was cylindrical in shape, and every place that was then known was situated on the flat end of the cylinder; and Plato, on the ground that the cube was the most perfect geometrical figure, imagined the earth to be a cube, the part of the earth known to the Greeks being on the upper surface. In these matters, indeed, the vaunted Greek mind was little in advance of the predecessors of the Vedic priests.

CHAPTER II.

THE FIRST GLIMPSES OF EGYPTIAN ASTRONOMY.

In the general survey, which occupied the preceding chapter, of the records left by the most ancient peoples, it was shown

THE ROSETTA STONE. (*In the British Museum.*)

that Egypt, if we consider her monuments, came first in the order of time. I have next to show that in the earliest monuments we have evidences of the existence and utilisation of astronomical knowledge.

It is impossible to approach such a subject as the astronomy of the ancient Egyptians without being struck with surprise that any knowledge is available to help us in our inquiries. A century ago, the man to whom we owe more than to all others in this matter; the man who read the riddle of those strange hieroglyphs, which, after having been buried in oblivion for nearly two thousand years, were then again occupying the learned, was not yet born. I refer to Champollion, who was born in 1790 and died in the prime of his manhood and in the midst of his work, in 1832.

Again, a century ago the French scientific expedition, planned by the great Napoleon, which collected for the use of all the world facts of importance connected with the sites, the buildings, the inscriptions, and everything which could be got at relating to the life and language of the ancient Egyptians, had not even been thought of; indeed, it only commenced its labours in 1798, and the intellectual world will for ever be a debtor to the man who planned it.

I know of no more striking proof of the wit of man than the gradual unravelling of the strange hieroglyphic signs in which the learning of the ancient Egyptians was enshrined; and there are few things more remarkable in the history of scientific investigation than the way in which a literature has been already brought together which is appalling in its extent; and yet it may well be that, vast as this literature is at present, it is but the vanguard of a much more stupendous one to follow; for we are dealing with a nation which we now know existed completely equipped in many ways at least seven thousand five hundred years ago.

It forms no part of the present work to give an account of

the unravelling to which I have referred, one which finds a counterpart in the results achieved by the spectroscope in another scientific field.

But a brief reference to one of the most brilliant achievements of the century may be permitted, and the more as it will indicate the importance of one of the most valued treasures in our national collections. I refer to the Rosetta Stone in the Egyptian Gallery of the British Museum. It was the finding of this stone in 1799 by Boussard, a captain of French artillery at Rosetta, which not only showed the baselessness of the systems of suggested interpretations of the hieroglyphics which had been in vogue from the time of Kircher downwards, but by its bilingual record in hieroglyphic, demotic and Greek characters, paved the way for men of genius like Thomas Young (1814) and Champollion (1822). The latter must be acknowledged as the real founder of the system of interpretation which has held its own against all opposition, and has opened the way to inquiries into the history of the past undreamt of when the century was young. Chateaubriand nobly said of him, "Ses admirables travaux auront la durée des monuments qu'il nous a fait connaître."

The germ of Champollion's discovery consisted in the bringing together of two sets of characters enclosed in cartouches. One of them is in the Rosetta inscription itself; the other, on the plinth of an obelisk in the island of Philæ. The name of Cleopatra was associated with the one inscription, and that of Ptolemy with the other. It was clear that if the two names, written [cartouche] and [cartouche], were really Ptolemaios and Cleopatra, they must include several identical signs or letters; in Ptolemaios the quadrangular figure □, being the first, must stand for *P*, and this in

Cleopatra was found to occur in the right place, standing fifth in order. The third sign in Ptolemaios must be an *o*, and the fourth an *l*. Now the lion for *l* occurs second in Cleopatra, and the knotted cord for *o* fourth. In this way, proceeding by comparison with other names, that of Alexander, or Alksantrs, was next discovered, and by degrees the whole Egyptian alphabet was recovered.

What had come down the stream of ages and were universally recognised as unsurpassed memorials of a mysterious past were the famous pyramids, successively described by Herodotus, Diodorus and Pliny among classical, and Abd el-Latîf among Arabian, chroniclers.

Although the rifling of the most important of these structures for the purpose of finding treasure dates at least as far back as 820 A.D., the Khalîf El-Mamun being the destroyer, the scientific study of their mode and objects of construction is a work of quite modern times, and may be said to have been inaugurated by Colonel Howard Vyse in 1839.

Much that has been written has been wild and nonsensical, but from the exact descriptions and measures now available, it is impossible to doubt that these structures *were erected by a people possessing much astronomical knowledge.* The exact orientation of the larger pyramids in the pyramid-field of Gîzeh has been completely established, and it is not impossible that some of the mysterious passages to be found in the pyramid of Cheops may have had an astronomical use.

Let us, to continue the subject-matter of the present chapter, come to the year 1820. It was about then that were gathered some of the first-fruits of the investigations carried on by the Commission to which I have referred; that some translations of

TEMPLE OF EDFÛ, LOOKING EAST: SHOWING PYLON AND OUTER COURT.

the inscriptions had been attempted, and that some of the new results were discussed by the members of the French Academy, while at the same time they astounded and delighted the outside world.

From the point of view which now concerns us, it may be said that the new discoveries might be arranged into three different groups. First of all, the land had been found full of temples, vast and majestic beyond imagination; among these the temples at Karnak were supreme, but there were others on a par with them in points of architectural detail. But besides these, then as now, above ground and inviting inspection, there were many others which were then—as undoubtedly many are still—more or less buried in the sand; some of these have since been unearthed to reveal the striking features of their structure.

I shall show subsequently that, on the evidence of the ancient Egyptians themselves, these temples were constructed in strict relation to stars; they, then, like the pyramids, must be taken as indicating astronomical knowledge.

If we deal with the general external appearance of the temples, they may be arranged architecturally into two main groups. Edfû is the most perfect example of the first group, characterised by having a pylon consisting of two massive structures right and left of the entrance, which are somewhat like the two towers that one sometimes sees on the west front of our English cathedrals.

In Denderah we have an example of the second group, in which the massive pylon is omitted. In these the front is entirely changed; instead of the pylon we have now an open front to the temple with columns—the Greek form of temple is approached.

Associated with many of the temples, frequently but not

GREAT COURT OF HEAVEN, AT THE ENTRANCE TO THE HATHOR TEMPLE AT DENDERAH.

universally in close proximity to the propylon, were obelisks, often of gigantic proportions, exceeding one hundred feet in height and many hundreds of tons in weight, which it has since been discovered were hewn out of the syenite quarry at Aswân, and floated down the river to the various places where they were to be erected.

It is not necessary to go to Egypt to see these wonderful monoliths, for they have been carried away from their original temple sites at Thebes and Heliopolis to adorn more modern cities in the Western world. London, Paris, Rome, and Constantinople are thus embellished. It is obvious to anyone

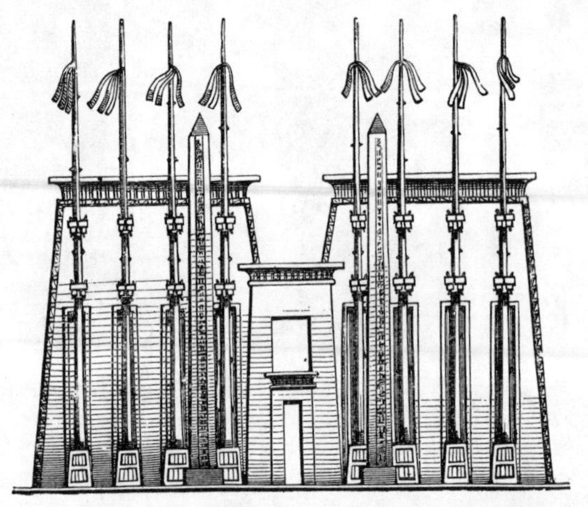

TEMPLE GATE WITH PROPYLON AND OBELISKS.

acquainted with astronomical history and methods, that some of these structures, at all events, may have served as gnomons.

Sometimes these temples, instead of being entirely constructed of stone on a level surface, were either entirely or partly rock-hewn. Of the former class, the temple of Abu Simbel is the most striking example; of the latter, the temple of Dêr el-Bahari at Thebes.

HATHOR TEMPLE OF DÊR EL-BAHARI. (*As restored by M. Brune.*)

The second revelation was that the walls of these temples, and of many funereal buildings, were, for the most part, covered with inscriptions in the language which was then but gradually emerging from the unknown, its very alphabet and syllabary being still incomplete. Hence there was not only a great wealth of temple buildings, but a still more wonderful wealth of temple inscriptions.

There was even more than this, and something more germane to our present purpose. In several temples which were examined, zodiacs—undoubted zodiacs, representing a

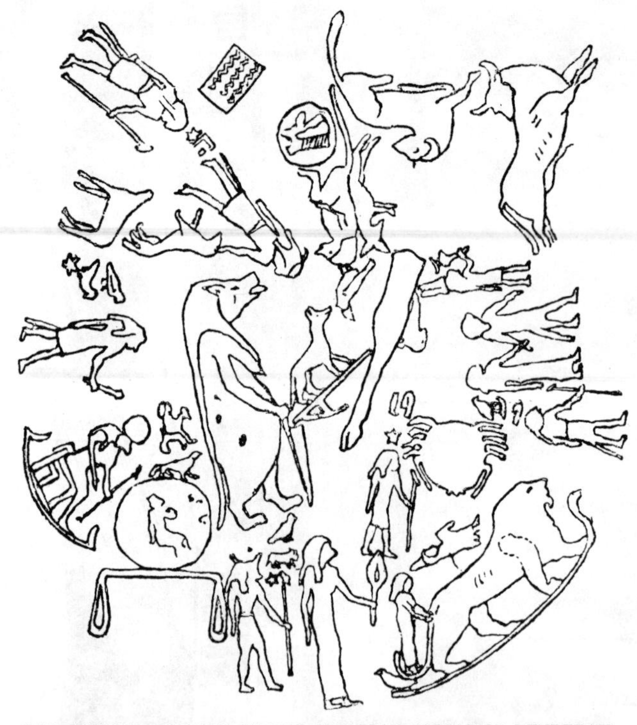

THE CENTRAL PORTION OF THE CIRCULAR ZODIAC OF DENDERAH.

third group of finds—were discovered; these, also, were accompanied by inscriptions of an obviously astronomical nature.

At the first blush, then, it seemed to be perfectly certain that we had to deal with a people of an astronomical turn of mind; and here was the opportunity for the astronomer, which indeed the French astronomers did not fail to make use of. Where the philologist was for the moment dumb, it seemed as if the astronomer could be of use, giving explanations, fixing probable dates on the one hand; while, on the other, he would certainly be gaining a fresh insight into, and possibly filling a tremendous gap in, the history of his science.

The figure on the preceding page gives an idea of the method of presentation generally employed in these zodiacs.

I shall show in the sequel—for I shall have to deal with this part of my subject at full length in a subsequent chapter—that many of the animal forms represent at once mythological personages and actual constellations.

CHAPTER III.

THE ASTRONOMICAL BASIS OF THE EGYPTIAN PANTHEON.

It will be abundantly clear from the statements made in the foregoing chapter that, as I have said, the main source of information touching things Egyptian consists no longer in writings like the Vedas, but in the inscriptions on the monuments, and the monuments themselves. It is true that, in addition to the monuments, we have the Book of the Dead, and certain records found in tombs; but, in the main, the source of information which has been most largely drawn upon consists in the monuments themselves—the zodiacs being included in that term.

It has been impossible, up to the present time, to fix with great accuracy the exact date of the earliest monuments. This should not surprise us. We must all feel that it is not a question of knowing so little—it is a question of knowing anything at all. When one considers that at the beginning of this century not a sign on any of these monuments was understood, and that now the wonderful genius of a small number of students has enabled Egyptologists to read the inscriptions with almost as much ease and certainty as we read our morning papers: *this* is what is surprising, and not the fact that we as yet know so little, and in many cases lack certainty.

But we already know that probably some of these monuments are nearly 6,000 years old. This has been determined by the convergence of many lines of evidence.

One of the many points already profoundly investigated by Egyptologists has been the chronology of the kings of Egypt

TABLET OF KINGS AT ABYDOS.

from their first monarch, whom all students recognise as Mena or Menes. All these inquirers have come to the definite conclusion that there was a King Mena, and that he reigned a long time ago; but with all their skill the final result is that they cannot agree to the date of this king within a thousand years; one reason among many others being that in these early days astronomy was a science still to be cultivated, and therefore the early Egyptians had not a perfect mode of recording; perhaps even they had no idea of a hundred years as we have. We are told that all their reckonings were the reckonings of the reigns of kings. This is difficult to believe, and the statement may be a measure of our ignorance of their method of record. We now, fortunately for us, have a calendar which enables us to deal with large intervals of time, but still we sometimes reckon, in Egyptian fashion, by the reigns of kings in our Acts of Parliament. Furthermore, Egypt being then a country liable to devastating wars, and to the temporary supremacy of different kingly tribes, it has been very difficult to disentangle the various lists of kings so as to obtain one chronological line, for the reasons that sometimes there were different kings reigning at the same time in different regions. The latest date for King Mena is, according to Bunsen, 3600 years B.C.; the earliest date, assigned by Boeckh, 5702 years B.C.; Unger, Brugsch, and Lepsius give, respectively, 5613, 4455, 3892. For our purpose we will call the date 4000 B.C. —that is, 6,000 years ago—and for the present consider this as the start-point for the long series of remains of various orders to which reference has been made, and from which alone information can be obtained.

We come now to deal with the ideas of the early inhabitants of the Nile valley. We find that in Egypt we are in presence absolutely of the worship of the Sun and of the accompanying

Dawn. Whatever be the date of the Indian ideas to which we have referred, we find them in Egypt in the earliest times. The ancient Egyptians, whether they were separate from, or more or less allied in their origin to, the early inhabitants of India, had exactly the same view of Nature-worship, and we find in their hymns and the lists of their gods that the Dawn and the Sunrise were the great revelations of Nature, and the things which were most important to man; *and therefore everything connected with the Sunrise and the Dawn was worshipped.*

HARPOCRATES.

Renouf, one of the latest writers on these subjects, says:[1] "I fear Egyptologists will soon be accused, like other persons, of seeing the dawn everywhere," and he quotes with approbation this passage from Max Müller relating to the Veda:—

> "I look upon the sunrise and sunset, on the daily return of day and night, on the battle between light and darkness, on the whole solar drama in all its details, that is acted every day, every month, every year, in heaven and in earth, as the principal subject."

But we must now go somewhat further into detail. The various apparent movements of the heavenly bodies which are produced by the rotation and the revolution of the earth, and the effects of precession, were familiar to the Egyptians, however ignorant they may have been of their causes; they carefully studied what they saw, and attempted to put their knowledge together in the most convenient fashion, associating it with their strange imaginings and their system of worship.

Dealing with the earth's rotation, how did the Egyptians

[1] Hibbert Lectures, 1879.

picture it? How was this interaction, so to speak, between the earth and the sky mythologically represented? They naturally would be familiar with the phenomena of dawn and sunset, more familiar certainly with the phenomenon of dawn than we are, because they had a climate much better suited for its study than ours. There can be no doubt that the wonderful

RĀ. MIN-RĀ. AMEN-RĀ.

scenes which they saw every morning and evening were the first things which impressed them, and they came to consider the earth as a god, surrounded by the sky—another god.

I have next to point out that, the sun being very generally worshipped in Egypt, there were various forms of the sun-god, depending upon the positions occupied in its daily course. We have the form of Harpocrates at its rising, the child sun-god being generally represented by the figure of a hawk. When in human form, we notice the presence of a side lock of hair. The god Rā symbolises, it is said, the sun in his noontide strength; while for the time of sunset we have various names, chiefly Osiris, Tum, or Atmu, the dying sun represented by a mummy and typifying old age. The hours of the day were also personified, the twelve changes during the twelve hours

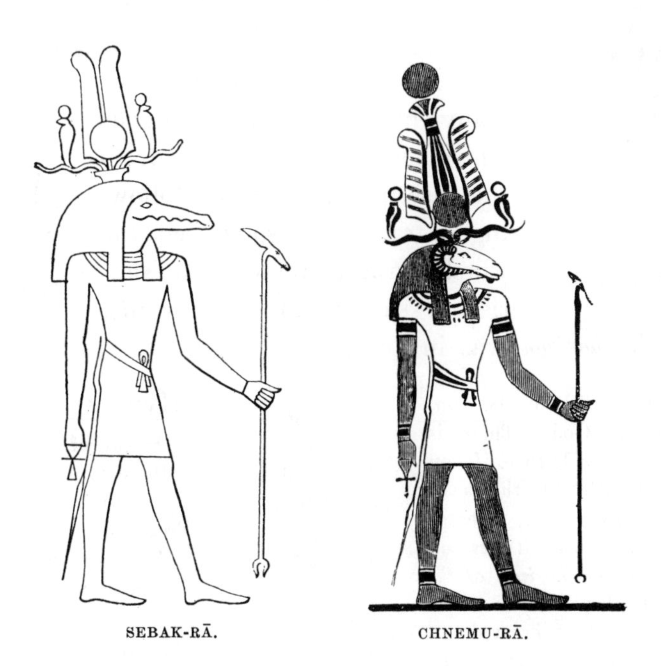

SEBAK-RĀ. CHNEMU-RĀ.

being mythically connected with the sun's daily movement across the sky.

We often find Rā compounded with other names, and in these forms of the god we possibly get references to the sun at different times of the year. Amen-Rā, Sebak-Rā, and Chnemu-Rā are cases in point. The former undoubtedly refers to the sun at the summer solstice. Min-Rā is an ithyphallic form.

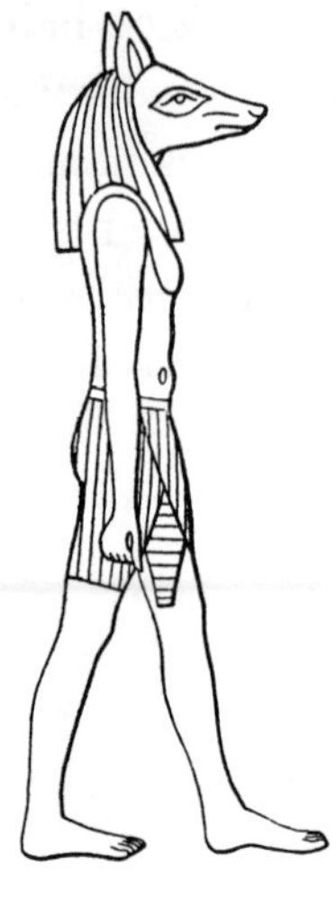
ANUBIS, OR SET.

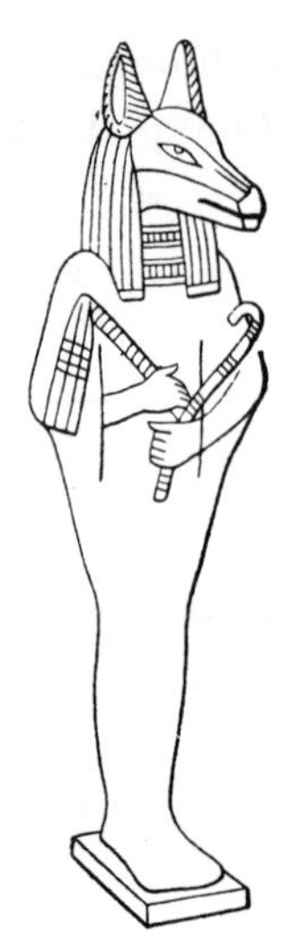
ANUBIS-OSIRIS.

The names given by the Egyptians to the sun then may be summarised as follows:—

Hor, or Horus, or Harpocrates, and Chepera (morning sun).
Rā (noon).
Tum or Atmu (evening sun).
Osiris (sun when set).

I have not space to quote the many hymns to the Sun-gods which have been recovered from the inscriptions, but the following extracts will show that the worship was in the main at sunrise or sunset—in other words, that the *horizon* was in question:—

"Thou disk of the Sun, thou living God! There is none other beside thee. Thou givest health to the eyes through thy beams, Creator of all beings. *Thou goest up on the eastern horizon of the heaven* to dispense life to all which thou hast created—to man, four-footed beasts, birds, and all manner of creeping things on the earth where they live. Thus they behold thee, and they go to sleep when thou settest."

Hymn to Tmu—

"Come to me, O thou Sun,
Horus of the horizon, give me help."

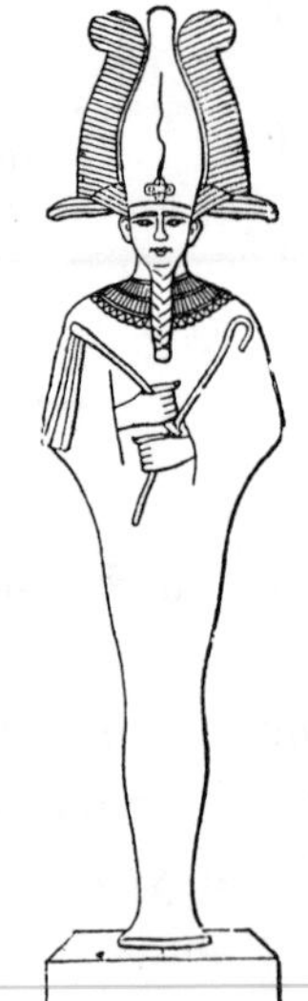
OSIRIS (AS A MUMMY).

Hymn to Horus—

"O Horus of the horizon, there is none other beside thee,
Protector of millions, deliverer of tens of thousands."

Hymn to Ra-Tmu-Horus—

"Hail to thee of the double horizon, the one god living by Maāt. . . . I am the maker of heaven and of the mysteries of the twofold horizon."

Hymn to Osiris—

"O Osiris! Thou art the

youth *at the horizon* of heaven daily, and thine old age at the beginning of all seasons. . . .

"The *ever-moving* stars are under obedience to him, and so are *the stars which set.*"

OSIRIS SEATED.

Hymn to Rā—

O Rā! in thine egg, radiant in thy disk, shining forth from the horizon, swimming over the steel (?) firmament.

"Tmu and Horus of the horizon pay homage to thee (Amen-Rā) in all their words."

So far we have dealt with the powers of sunlight; but the ancient Egyptians, like ourselves, were familiar with the powers of darkness or of the underworld. The chief god antithetical to the sun was variously named—Sit, Set, Sut,

Anubis, Typhon, Bes; and a host of other names was given to him. As I shall show, the idea of darkness was associated with the existence of those stars which never set, so that even here the symbolism was astronomical.

VARIOUS FORMS OF BES—AS WARRIOR, MUSICIAN, AND BUFFOON.

The contrast between the representations of Bes and of the other forms suggests that the former was imported. In the form of Typhon the goddess Taurt is represented as a hippopotamus, while for Anubis the emblem is a jackal.

In all illustrations of funeral ceremonies the above-mentioned figure largely. In the Book of the Dead we find that in the representations of the judgment of the dead, besides Osiris we have Anubis, both responsible for the weighing of the soul.

With the moon we find two gods connected—Thoth-lunus and Khons-lunus—though the connection is not a very obvious one.

Thoth is also associated with the Egyptian year, and is variously represented; all forms, however, are based upon the ibis.

For the stars generally we find a special goddess, Sesheta.

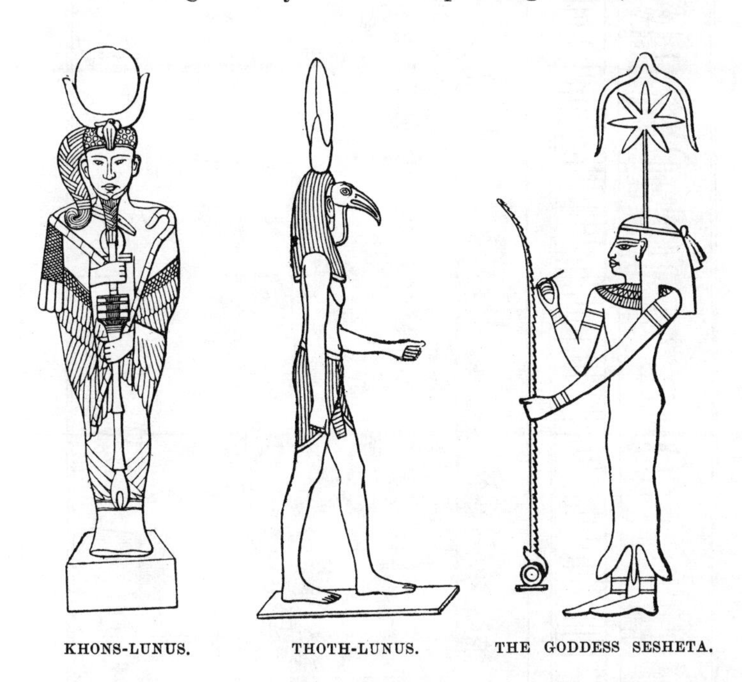

KHONS-LUNUS. THOTH-LUNUS. THE GODDESS SESHETA.

Thoth as the sacred scribe and Sesheta as the star-goddess are often represented together engaged in writing.

Associated with the phenomena of morning and evening we find the following divinities. The attributes stated are those now generally accepted. This is a subject which will occupy us in the sequel.

Isis represents the Dawn and the Twilight; she prepares the way for the Sun-god. The rising sun between Isis and Nephthys = morning.

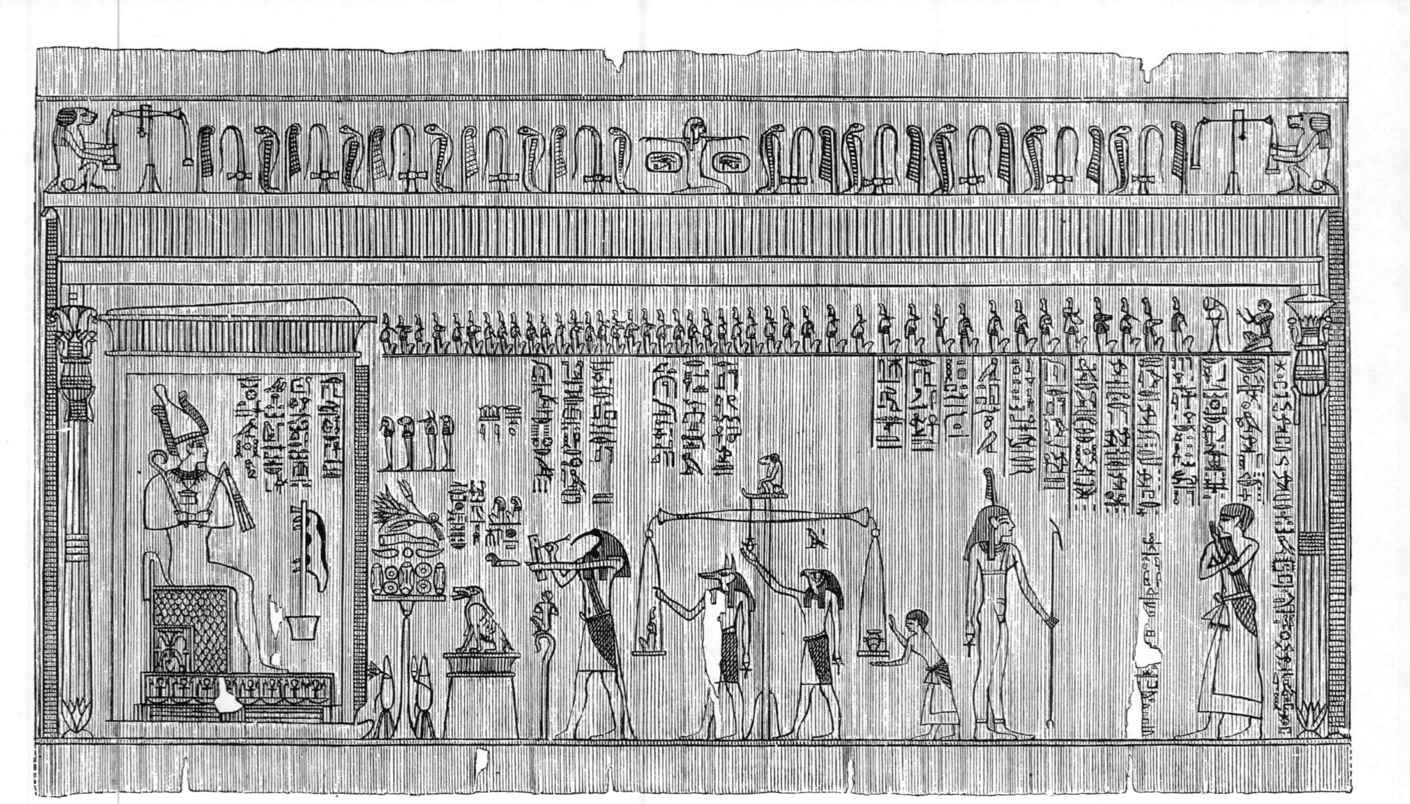

THE WEIGHING OF THE SOUL BY HORUS AND ANUBIS, IN PRESENCE OF OSIRIS.

Nephthys is the Dawn and the Twilight, sometimes Sunset.

Shu is also the Dawn, or sunlight. Tefnut represents the coloured rays at dawn. Shu and Tefnut are the eyes of Horus. Shu was also called "Neshem," which means green felspar, in consequence of the green colour observed at dawn. The green tint at dawn and sunset are represented further by the "sycamore of emerald." Sechet is another goddess of the Dawn, the fiery Dawn.

THOTH AND SESHETA WRITING THE NAME OF RAMESES II. ON THE FRUIT OF THE PERSEA.

(*Relief from the Ramesseum at Thebes.*)

The *red* colours at sunset were said to be caused by the blood flowing from the Sun-god when he hastens to his suicide. A legend describes Isis as stanching the blood flowing from the wound inflicted on Horus by Set.

CLEOPATRA AS THE GODDESS ISIS.

Hathor is, according to Budge, identified with Nu or Nu-t, the sky, or place in which she brought forth and suckled Horus. She is the female power of Nature, and has some of the attributes of Isis, Nu-t, and Maāt.

We next have to gain some general idea of the Egyptian

cosmogony—the relation of the sun and dawn to the sky; this is very different from the Indian view. The Sky is Nu or Nu-t, represented as a female figure bending over Seb, the Earth, with her feet on one horizon and her finger-tips on the other. Seb is represented by a recumbent figure,

ISIS (SEATED).

while the sky, represented by the goddess Nu-t, is separated from the earth by Shu, the god of air or sunlight. The daily journey of the sun is represented by a god in a boat traversing the sky from east to west. The goddess Nu-t is variously symbolised. Sometimes there is a line of stars along her back, which clearly defines her nature, but sometimes she is represented by a figure in which the band of stars is accompanied by a band of water. This suggests the Jewish idea of the firmament. We read of the firmament in the midst of the waters, which divided the waters from the waters, the

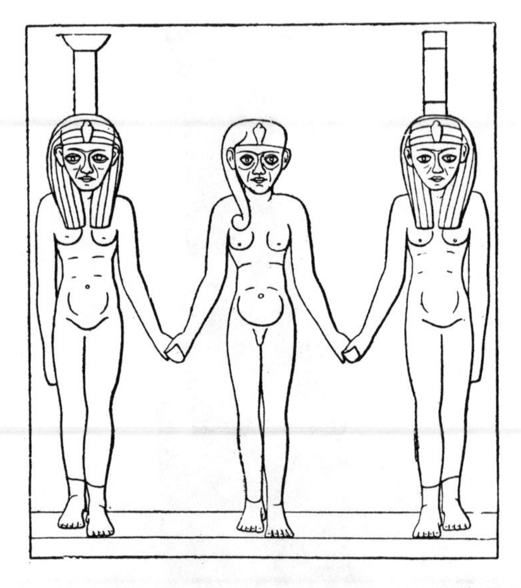

THE RISING SUN HORUS BETWEEN ISIS AND NEPHTHYS.

waters above being separated from the waters below the firmament.

It would seem that it was not very long before the Egyptians saw that the paths of the sun and stars above the horizon were extremely unequal: in the case of the sun, at different times of the year; in the case of the stars, depending upon their position near the equator or either pole. In this way, perhaps, we may explain a curious variant of the drawing of the goddess Nu-t, in which she is represented double, a larger one stretching over a smaller one.

Not only the Sun-gods, but the stars, were supposed to travel in boats across the firmament from one horizon to the other. The under-world was the abode of the dead; and daily the sun, and the stars which set, died on passing to the regions of the west, or Amenti, below the western horizon, to be born again on the eastern horizon on the morrow.

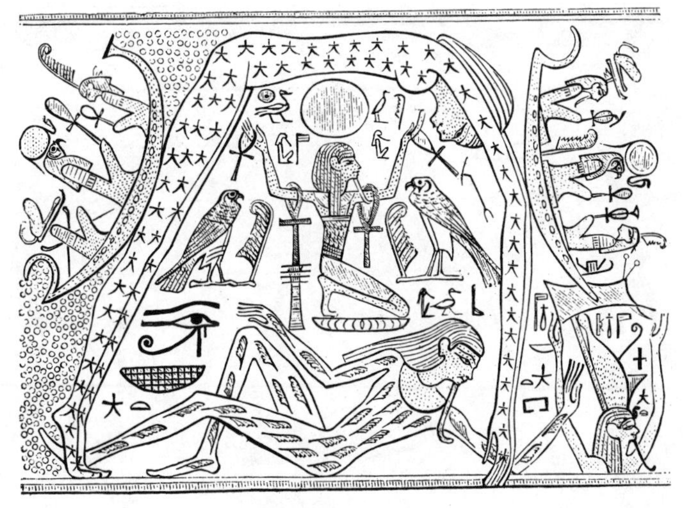

THE GODDESS NU-T.

In this we have the germ of the Egyptian idea of immortality.

Among other gods which mav be mentioned are Chnemu, the "Moulder," who was thought to possess some of the attributes of Rā; and Ptah, the "Opener," who is at times represented with Isis and Nephthys, and then appears as a form of Osiris.

We can now begin to glimpse the Egyptian mythology.

Seb, the Earth, was the husband of Nu-t, the Sky; and

the Sun- and Dawn-gods and -goddesses were their children, as also were Shu representing sunlight, and Tefnut representing the flames of dawn.

Maāt, the goddess of law, was the daughter of Rā.

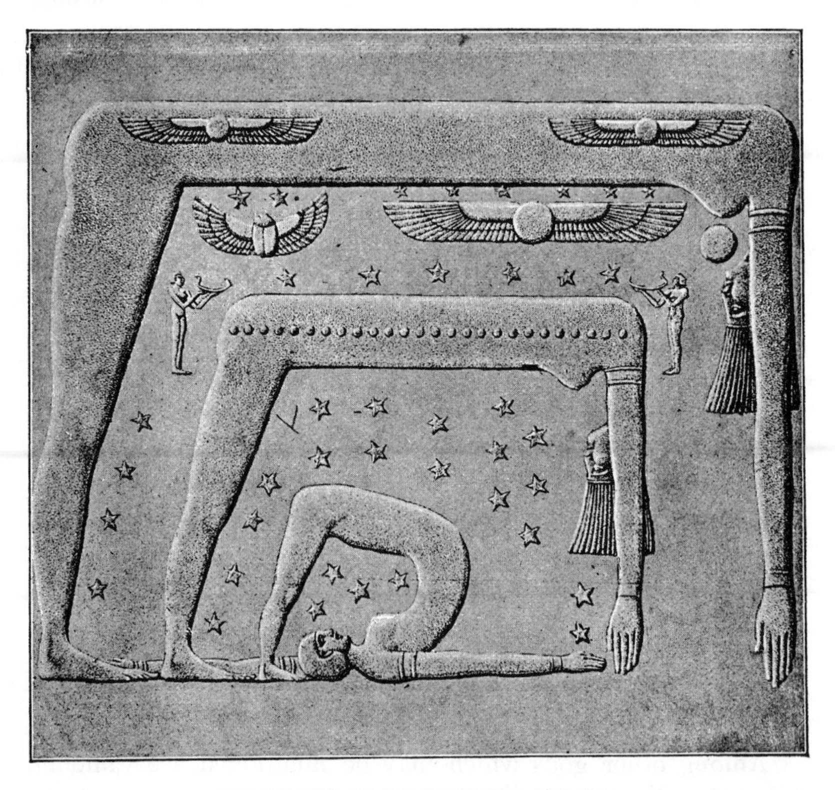

THE GODDESS NU-T REPRESENTED DOUBLE.

We know several points regarding Egyptian customs independently of the astronomical inscriptions, properly so-called, to which I have called attention. We know that there were sacrifices at daybreak; we know that stars were watched before sunrise, and heralded the dawn; we know that these observations were among the chief duties of the sacrificial priests, and it is obvious that a knowledge of star-places, as well as star-names, must have been imperative to these morning watchers, who eventually compiled lists of decans—that is, lists of belts of stars extending round the heavens, the risings of which followed each other by ten days or so.

VARIOUS FORMS OF SHU.

These are the exact equivalents of the moon-stations which the Indians, Arabians, and other peoples invented for the same purpose. We also find, more or less indeterminately from inscriptions in some graves at Thebes, that the daily risings of the chief stars were observed very carefully throughout the year. Unfortunately the inscriptions in question are very difficult indeed to co-ordinate. There have been various efforts made to connect them with certain stars, but, so far, I am afraid they have resisted all efforts to get a complete story out of them, though certain very important points have been made out. These points I shall consider later.

It is not too early to point out here that there is evidence

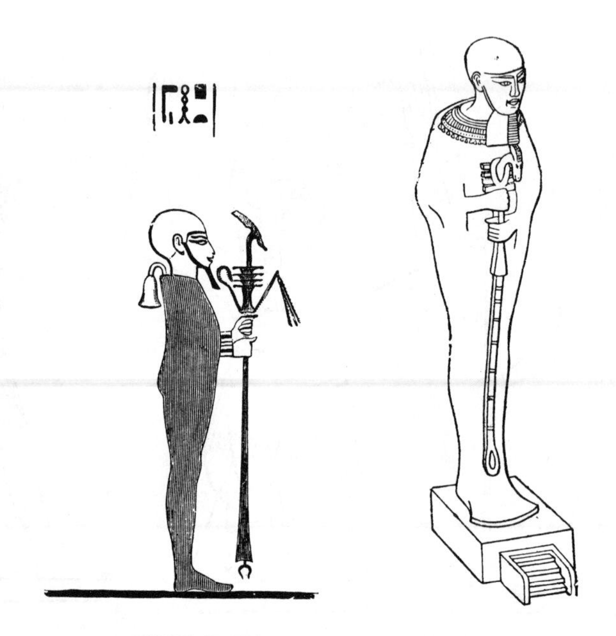

FORMS OF PTAH, THE GOD OF MEMPHIS.

that the Egyptian pantheon, as I have stated it, had not a simple origin. There are traditions that many of the gods came from a region indeterminately described as the land of Punt. Among these gods are Chnemu, Amen-Rā, Hathor, and Bes. On page 28 I have associated Bes with Typhon, following several authorities, but if they are right it is very difficult to understand his *rôle*. It may also be added that the temple-evidence supports the view of his foreign origin.[1]

When one comes to consider the Rig-Veda and the Egyptian monuments from an astronomical point of view, one is struck by the fact that, in both, the early worship and all the early observations related to the horizon. This was true not only of the sun, with which so far we have exclusively dealt, but it was equally true of the stars which studded the general expanse of sky.

In Egypt, then, as in India, the pantheon was astronomical and, to a very large extent, solar in origin. I shall have to show that the remainder—nearly the whole of it—had its origin in stellar relations.

[1] *See* Rawlinson's "History of Egypt," Vol. II., p. 134, for references on this subject.

CHAPTER IV.

THE TWO HORIZONS.

It is not only of the first importance for our subject, but of great interest in itself, to study some of the astronomical problems connected with this horizon worship, which in the previous chapter we have found to be common to the early peoples of India and Egypt.

We must be perfectly clear before we go further what

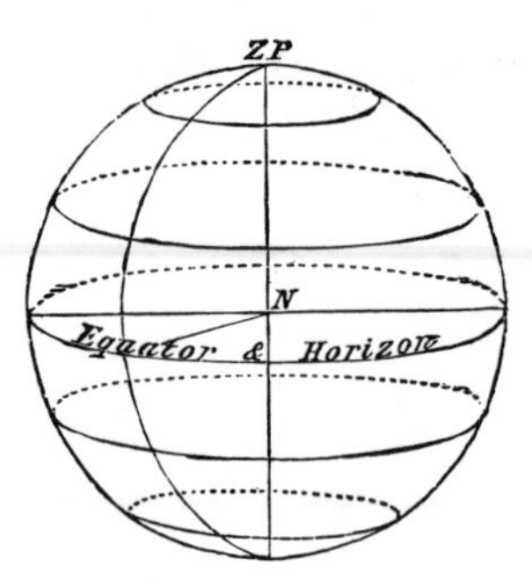

APPARENT MOVEMENT OF THE STARS TO AN OBSERVER AT THE NORTH POLE.

this horizon really is, and for this some diagrams are necessary.

The horizon of any place is the circle which bounds our view of the earth's surface, along which the land (or sea) and sky appear to meet. We have to consider the relation of the horizon of any place to the apparent movements of celestial bodies at that place.

We know, by means of the demonstration afforded by Foucault's pendulum, that the earth rotates on its axis, but this idea was, of course, quite foreign to these early peoples. Since the earth rotates with stars, infinitely removed, surrounding it on all sides, the apparent movements of the stars will depend very much upon the position we happen to occupy on the earth: this can be made quite clear by a few diagrams.

An observer at the North Pole of the earth, for instance, would see the stars moving round in circles parallel to the horizon. No star would either rise or set—one half of the heavens would be always visible above his horizon, and the other half invisible; whereas an observer at the South Pole would see that half of the stars invisible to the observer at

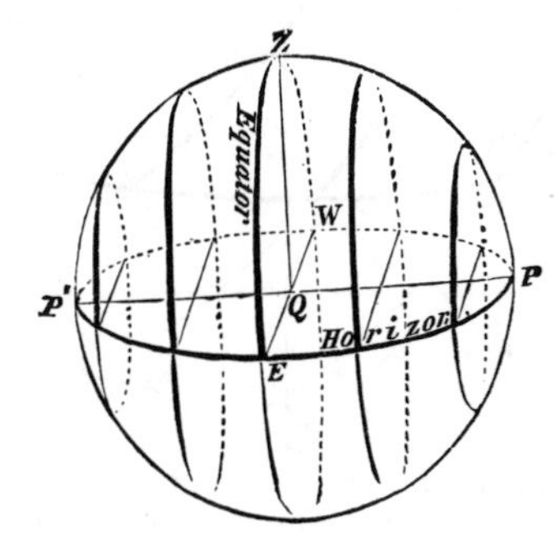

APPARENT MOVEMENT OF THE STARS TO AN OBSERVER AT THE EQUATOR.

the northern one, because it was the half below the N. horizon. If the observer be on the equator, the movements of the stars all appear as indicated in the above diagram—that is, all the stars will rise and set, and each star in turn will be half its time above the horizon, and half its time below it. But if we consider the position of an observer in middle latitude, say in London, we find that some stars will always be above the horizon, some always below—that is, they will neither rise nor set. All other stars will both rise and set, but some of them will be above the horizon for a long time and below for a short time, whereas others will be a very short time above the horizon and a long time below it.

At *O* we imagine an observer to be in latitude 45° (that is, half-way between the equator in latitude 0°, and the North Pole in latitude 90°), hence the North Celestial Pole will be half-way between the zenith and the horizon; and close to the pole he will see the stars describing circles, inclined, however, and not retaining the same distance from the horizon. As the

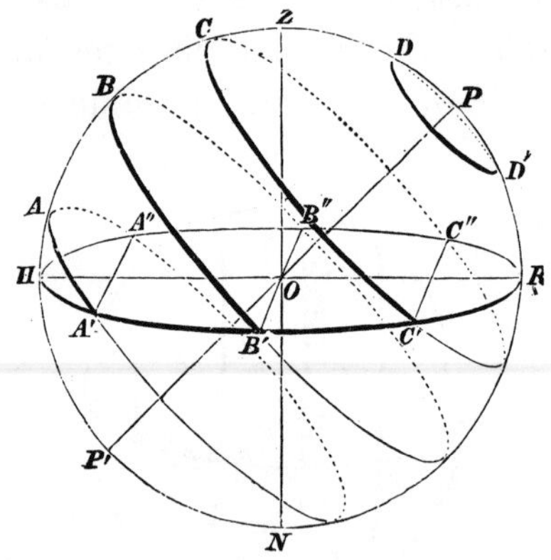

THE CELESTIAL SPHERE VIEWED FROM A MIDDLE LATITUDE.
AN OBLIQUE SPHERE.

eye leaves the pole, the stars rise and set obliquely, describe larger circles, gradually dipping more and more under the horizon, until, when the celestial equator is reached, half their journey is performed below it. Still going south, we find the stars rising less and less above the horizon, until, as there were northern stars that never dip below the horizon, so there are southern stars which never appear above it. *D D'* shows the apparent path of a circumpolar star; *B B' B''* the path and rising and setting points of an equatorial star; *C C' C''* and *A A' A''* those of stars of mid-declination, one north and the other south.

Wherever we are upon the earth we always imagine that we are on the top of it. The idea held by all the early peoples was that the earth was an extended plain: they imagined that the land that they knew and just the surrounding lands were really in the centre of the extended plain. Plato, for instance, as we have seen, was content to

A TERRESTRIAL GLOBE WITH WAFER ATTACHED TO SHOW THE VARYING CONDITIONS OF OBSERVATION IN A MIDDLE LATITUDE.

put the Mediterranean and Greece upon the top of his cube, and Anaximander placed the same region at the top of his cylinder.

We can very conveniently study the conditions of observation at the poles of the earth, the equator, and some place in middle latitude, by using an ordinary terrestrial globe.

The wooden horizon of the globe is parallel to the horizon of a place at the top of the globe, which horizon we can represent by a wafer. In this way we can get a very concrete idea of the different relations of the observer's horizon in different latitudes to the apparent paths of the stars.

We have next to deal with the astronomical relations of the horizon of any place in connection with the worship of the sun and stars at the times of rising or setting, when, of course, they are on or near the horizon; and in order to bring this matter nearer to the ancient monuments, it will be convenient to study this question for Thebes, where they exist in greatest number and have been most accurately described.

To adjust things properly we must rectify the globe to the latitude of 25° 40′ N., or, in other words, incline the axis of the globe at that angle to the wooden horizon.

It will be at once seen that the inclination of the axis to the horizon is very much less than in the case of London. Since all the stars which pass between the North Pole and the horizon cannot set, all their apparent movement will take place above the horizon. All the stars between the horizon and the South Pole will never rise. Hence, stars within the distance of 25° from the North Pole will never set at Thebes, and those stars within 25° of the South Pole will never be visible there. At any place the latitude and the elevation of the pole are the same. It so happens that all these places with which archæologists have to do in studying the history of early peoples, Egypt, Babylonia, Assyria, China, Greece, &c., are in middle latitudes, therefore we have to deal with bodies in the skies, which do set, and bodies which do not; and the elevation of the pole is neither very great nor very small. In each different latitude the inclination of the equator to the horizon, as well as the elevation of the pole, will vary, but there will be a strict relationship between the inclination of the equator at each place and the elevation of the pole. Except at the poles themselves the equator will cut the horizon due east and due west. Therefore every celestial body which rises or sets to the north of the equator will cut the horizon between the east or west point and the north point; those bodies which do not set will, of course, not cut the horizon at all.

The sun, and stars near the equator, in such a latitude as that of Thebes, will appear to rise or set at no very considerable angle to the vertical; but when we deal with stars rising or setting near to the north or south points of the horizon they will seem to skim along the horizon instead of rising or setting vertically.

Now it will at once be obvious that there must be a strict law connecting the position of the sun (or a star) with its place of rising or setting. Stars at the same distance from either of the celestial poles will rise or set at the same point of the horizon, and if a star does not change its place in the heavens it will always rise or set in the same place.

Here it will be convenient to introduce one or two technical terms. Every celestial body, whether we deal with the sun, moon, planet, or star, occupies at any moment a certain place in the sky, partly, though not wholly, defined by what we term its declination, *i.e.*, its distance from the celestial equator. This declination is one of the two coordinates which are essential for enabling us to state accurately the position of any body on the celestial vault; and we must quite understand that if all these bodies rise and set, and rise and set visibly, the place of their rising or setting must be very closely connected with their declination. Bodies with the same declination will rise at the same points of the horizon. When the declination changes, of course the body will rise and set in different points of the horizon.

Next we define points on the horizon by dividing the whole circumference into four quadrants of 90° each=360°, so that we can have *azimuths* of 90° from the north or south points to the east and west points.

Azimuths are not always reckoned in this way, navigators preferring one method, while astronomers prefer another. Thus azimuth may also be taken as the distance measured in degrees from the south point in a direction passing through the

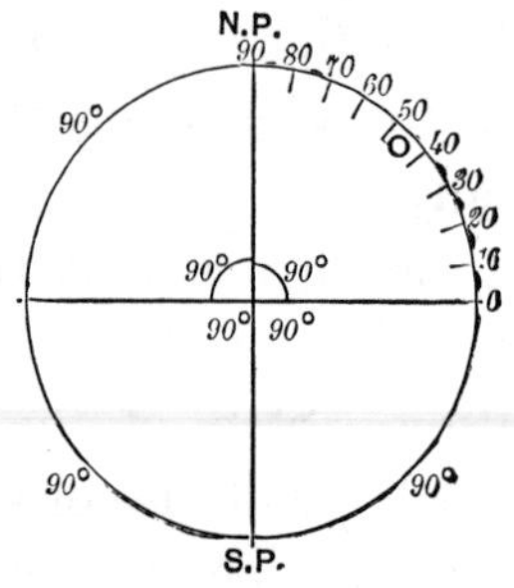

SHOWING AMPLITUDES RECKONED FROM THE EAST OR WEST POINTS TO N.P., NORTH POINT OF HORIZON, AND S.P., SOUTH POINT OF HORIZON.

west, north, and east points. On this system, a point can have an azimuth varying from 0° to 360°.

It is next important to define the term *amplitude*. The amplitude of a body on the horizon is its distance north and south of the east and west points; it is always measured to the nearest of these two latter points, so that its greatest value can never exceed 90°. For instance, the south point itself would have an amplitude of 90° south of west (generally written W. 90° S.), or 90° south of east (E. 90° S.), while a point 2° to the westward of south would have an amplitude of W. 88° S., and not E. 92° S.

We can say then that a star of a certain declination will rise or set at such an *azimuth*, if we reckon from the N. point of the horizon, or at such an amplitude if we reckon from the *equator*. This will apply to both north and south declinations.

The following table gives for Thebes the amplitudes of rising or setting (north or south) of celestial bodies having declinations from 0° to 64°; bodies with higher declinations than 64° never set at Thebes if they are north, or never rise if they are south, as the latitude (and therefore the elevation of the pole) there is nearly 26°.

AMPLITUDES AT THEBES.

Declination.	Amplitude.		Declination.	Amplitude.		Declination.	Amplitude.	
°	°	′	°	°		°	°	′
0	0	0	22	24	33	44	50	25
1	1	7	23	25	41	45	51	41
2	2	13	24	26	49	46	52	57
3	3	20	25	27	58	47	54	14
4	4	26	26	29	6	48	55	32
5	5	33	27	30	15	49	56	51
6	6	40	28	31	23	50	58	12
7	7	47	29	32	32	51	59	34
8	8	53	30	33	41	52	60	58
9	9	59	31	34	51	53	62	23
10	11	6	32	36	1	54	63	51
11	12	13	33	37	11	55	65	21
12	13	20	34	38	21	56	66	54
13	14	27	35	39	31	57	68	31
14	15	34	36	40	42	58	70	12
15	16	41	37	41	53	59	71	59
16	17	49	38	43	5	60	73	55
17	18	56	39	44	17	61	76	1
18	20	3	40	45	30	62	78	25
19	21	10	41	46	43	63	81	19
20	22	17	42	47	56	64	85	42
21	23	25	43	49	10			

The absolute connection, then, between the declination of a heavenly body and the amplitude at which it rises and sets is obvious from the above table: given the declination we know the amplitude; given the amplitude we know the declination.

Suppose we were dealing with a sea horizon: all the bodies rising or setting at the same instant of time would be in a great circle round the heavens, for the plane of the sensible horizon is parallel to the geocentric one.

But there are some additional points to be borne in mind. Ordinarily we should determine that the amplitude being so and so, the declination of the body which rose or set with that amplitude would be so and so, taking the horizon to be an all-round horizon like a sea one. But that would not be quite true, because we generally see the sun, to take an instance, some little time before it really rises and after it has set, owing to refraction. So that if we see the sun setting, say, north of west, we know that when we see it setting it appears really a little further to the north than it actually was at the moment of true sunset, because refraction gives us the position of the sun just below the true horizon. That is one point that we have to consider. Another is that, of course, we as a rule do not deal with sea horizons. Here we find a hill, there some other obstacle; so that it is necessary to make a correction depending on the height of the hill or other obstacle above the sea- or true-horizon at the place. Only when we take these things completely into consideration, can we determine absolutely the declination, or distance from the celestial equator, of the body at the moment of rising or setting. Still, it is worth while noting that when only approximations are required, the refraction- and hill-corrections have a tendency to neutralise each other in the northern hemisphere. Refraction will tend to carry the sunrise or sunset place more to the north,

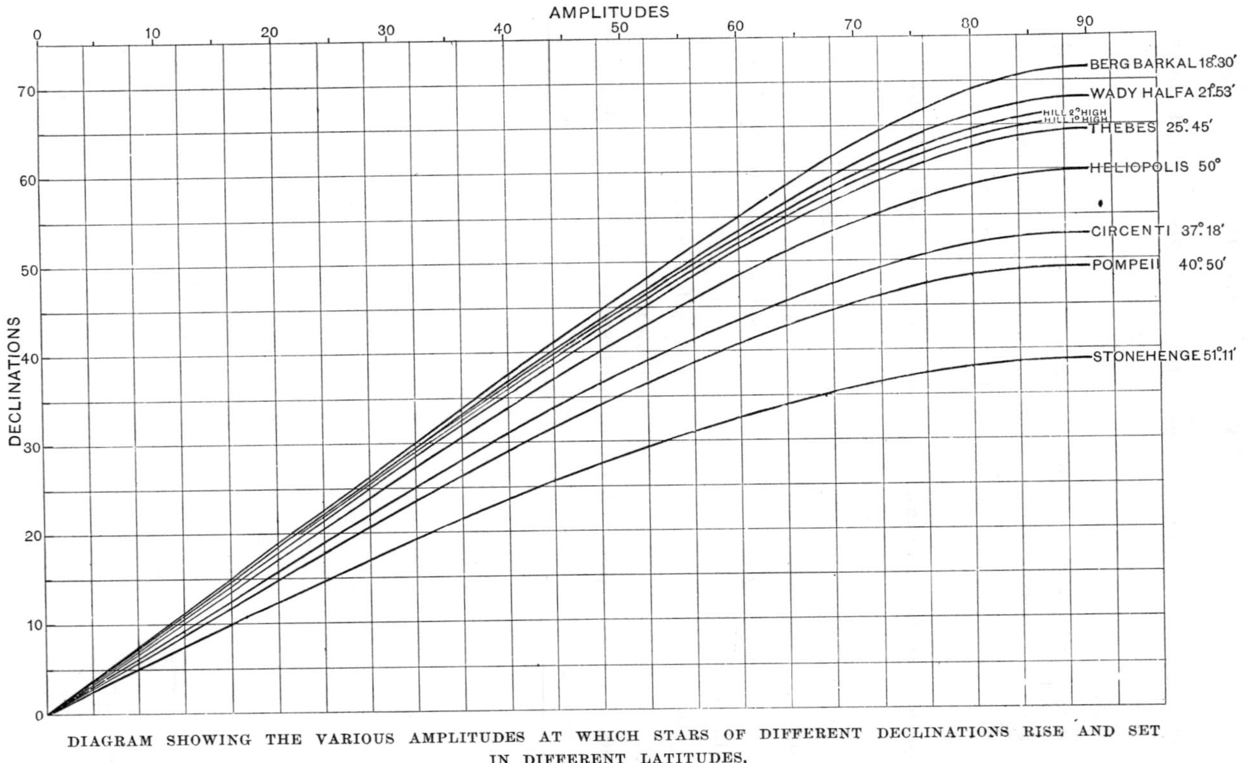

DIAGRAM SHOWING THE VARIOUS AMPLITUDES AT WHICH STARS OF DIFFERENT DECLINATIONS RISE AND SET IN DIFFERENT LATITUDES.

hills will cause the body to appear to rise or set more to the south.

It is important to point out that these corrections vary very considerably in importance according to the declination of the star with which we have to deal. With a high north or south declination the amplitude increases very rapidly, and the more it increases the more the corrections for refraction and elevation above the true horizon to which I have referred become of importance. In all cases the correction has to be made so that the amplitude will be increased or decreased from the true amplitude by this effect of refraction, according as the body—whether sun or star—is seen to the north or south of the equator.

In the diagram given on page 49, the various amplitudes are shown at which bodies of different declinations appear to rise and set in places with latitudes ranging from 19° to 51° N. It is a diagram to which frequent reference will be made in the sequel.

CHAPTER V.

THE YEARLY PATH OF THE SUN-GOD.

LET us, then, imagine the ancient Egyptians, furnished with the natural astronomical circle which is provided whenever there is an extended plain, engaged in their worship at sunrise, praying to the "Lord of the two Horizons." The rising (and setting) of stars we will consider later; it is best to begin with those observations about which there is the least question.

In the very early observations that were made in Egypt and Babylonia, when the sun was considered to be a god who every morning got into his boat and floated across space, there was no particular reason for considering the amplitude at which the supposed boat left or approached the horizon. But a few centuries showed that this rising or setting of the sun in widely varying amplitudes at different parts of the year depended upon a very definite law. We now, more fortunate than the early Egyptians, of course know exactly what this law is, and with a view of following their early attempts to grapple with the difficulties presented to them we must pass to the yearly path of the sun, in order to study the relation of the various points of the horizon occupied by the sun at different times in the year.

Not many years ago Foucault gave us a means of demonstrating the fact that the earth rotates on its axis. We have also a perfect method of demonstrating that the earth not only rotates on its axis once a day, but that it moves round the sun once a year, an idea which was undreamt of by the ancients. As a pendulum shows us the rotation, so the determination of the aberration of light demonstrates for us the revolution of the earth round the sun.

We have, then, the earth endowed with these two movements—a rotation on its axis in a day, and a revolution round the sun in a year. To see the full bearing of this on our present inquiry, we must for a time return to the globe or model of the earth.

To determine the position of any place on the earth's surface we say that it is so many degrees distant from the equator, and also so many degrees distant from the longitude of Greenwich: we have two rectangular co-ordinates, latitude and longitude. When we conceive the earth's equator extended to the heavens, we have a means of determining the positions of stars in the heavens exactly similar to the means we have of determining the position of any place on the earth. We have already defined distance from the equator as north or south declination in the case of a star, as we have north latitude or south latitude in case of a place on the earth. With regard to the other co-ordinate, we can also say that the heavenly body whose place we are anxious to determine is at a certain distance from our first point of measurement, whatever that may be, along the celestial equator. Speaking of heavenly bodies, we call this distance right ascension; dealing with matters earthy, we measure from the meridian of Greenwich and call the distance longitude.

The movement of the earth round the sun is in a plane which is called the plane of the ecliptic, and the axis of rotation of the earth is inclined to that plane at an angle of something like $23\frac{1}{2}^{\circ}$. We can if we choose use the plane of the ecliptic to define the positions of the stars as we use the plane of the earth's equator. In that case we talk of distance from the ecliptic as celestial latitude, and along the ecliptic from one of the points where it cuts the celestial equator as celestial longitude. The equator, then, cuts the ecliptic at two points: one of these is chosen for the start-point of measurement along both the equator and the ecliptic, and is called the first point of Aries.

We have, then, two systems of co-ordinates, by each of which we can define the position of the sun or a star in the heavens: equatorial co-ordinates dealing with the earth's equator, ecliptic co-ordinates dealing with the earth's orbit. Knowing that the earth moves round the sun once a year, the year to us moderns is defined with the most absolute accuracy. In fact, we have three years: we have a sidereal year—that is, the time taken by the earth to go through exactly 360° of longitude; we have what is called the tropical year, which indicates the time taken by the earth to go through not quite 360°, to go from the first point of Aries till she meets it again; and since the equinoctial point advances to meet the earth, we talk about the precession of the equinoxes; this year is the sidereal year minus twenty minutes. Then there is also another year called the anomalistic year, which depends upon the movement of the point in the earth's orbit where the earth is nearest to the sun; this is running away, so to speak, from the first point of Aries, instead of advancing to meet it, so that in this case we get the sidereal year plus nearly five minutes.

The angle of the inclination of the earth's plane of rotation to the plane of its revolution round the sun, which, as I have said, is at the present time something like $23\frac{1}{2}°$, is called *the obliquity of the ecliptic.* This obliquity is subject to a slight change, to which I shall refer in a subsequent chapter.

In order to give a concrete idea of the most important points in the yearly path of the earth round the sun, let us imagine four globes arranged on a circle representing the earth at different points of its orbit, with another globe in the centre representing the sun, marking the two practically opposite points of the earth's orbit, in which the axis is not inclined to or from the sun but is at right angles to the line joining the earth in these two positions, and the two opposite and intermediate points at which the north pole of the axis is most inclined towards and away from the sun.

A diagram will show what will happen under these conditions. If we take first the points at which the axis, instead of being inclined towards the sun, is inclined at right angles to it, it is perfectly obvious that we shall get a condition of things in which the movement of the earth on its axis will cause the dark side of the earth and also the light side represented by the side nearest to the sun, both being of equal areas, to extend from pole to pole; so that any place on the earth rotating under those conditions will be brought for half a period of rotation into the sunlight, and be carried for half a period of the rotation out of the sunlight; the day, therefore, will be of the same length as the night, and the days and nights will therefore be equal all over the world.

We call this the time of the equinoxes; the nights are of the

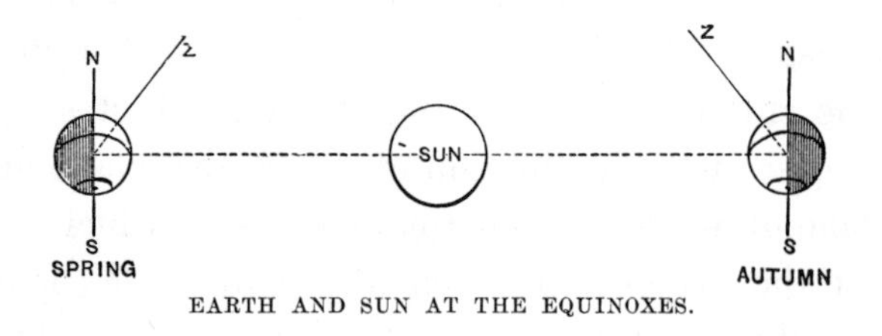

EARTH AND SUN AT THE EQUINOXES.

same length as the day in both these positions of the earth with regard to the sun.

In the next figure we have the other condition. Here the earth's axis is inclined at the greatest angle of $23°\frac{1}{2}$, towards, and away from, the sun. If I take a point very near the north pole, that point will not, in summer, be carried by the earth's rotation out of the light, and a part equally near the south pole will not be able to get into it. These are the conditions at and near two other points called the solstices.

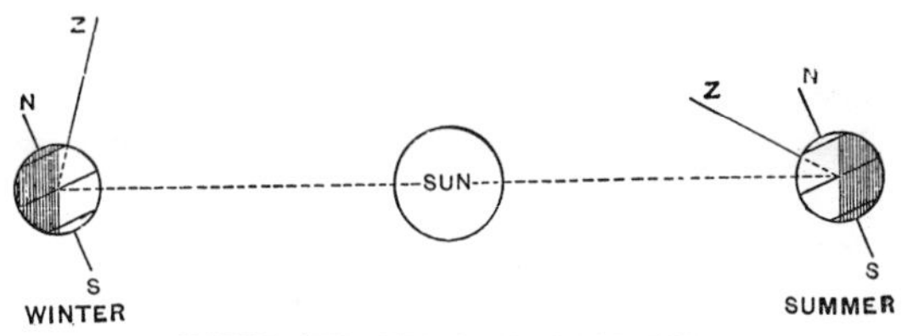

EARTH AND SUN AT THE SOLSTICES.

On each of these globes I have drawn a line representing the overhead direction from London. If we observe the angle between the direction of the zenith and that to the sun in winter we find it considerable; but if we take the opposite six-monthly condition we get a small angle.

In other words, under the first condition the sun at noon will be far from the zenith of London, we shall have winter; and in the other condition the sun will be as near as it can be to the zenith at noon, we shall have summer. These two cases represent the two points in the earth's orbit at which the sun has the greatest declination south and north. With the greatest north declination the sun will come up high, appear to remain at the same height above the horizon at noon for a day or two, as it does at our summer solstice, and then go down again; at the other point, when it has the greatest southern declination, it will go down to the lowest point, as it does in our winter, stop, and come up again—that is, the sun will stand still, so far as its height above the horizon at noon is concerned, and the Latin word solstice exactly expresses that idea. We have, then, two opposite points in the revolution of the earth round the sun at which we have equal altitudes of the sun at noon, two others when the altitude is greatest and least.

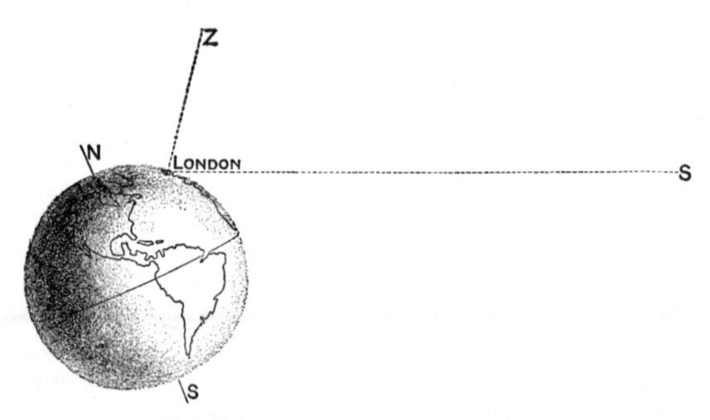

DIAGRAM SHOWING POSITION OF THE SUN IN RELATION TO THE ZENITH OF LONDON AT THE NORTHERN WINTER SOLSTICE.

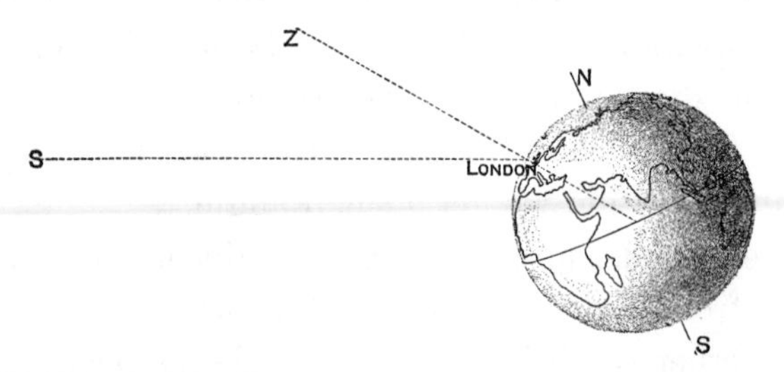

DIAGRAM SHOWING POSITION OF THE SUN IN RELATION TO THE ZENITH OF LONDON AT THE NORTHERN SUMMER SOLSTICE.

We get the equal altitudes at the equinoxes, and the greatest and the least at the solstices.

These altitudes depend upon the change of the sun's declination. The change of declination will affect the azimuth and amplitude of the sun's rising and setting; this is why, in our northern hemisphere, the sun rises and sets most to the north in summer and most to the south in winter. At the equinoxes the sun has always 0° Decl., so it rises and sets due east and west all over the world. But at the solstices it has its greatest declination of $23\frac{1}{2}$° N. or S.; it will rise and set, therefore, far from the east and west points; how far, will depend upon the latitude of the place we consider. The following are approximate values:

Latitude of Place.	Amplitude of Sun at Solstice.	
°	°	′
25	26	5
30	27	24
35	29	8
40	31	21
45	34	40
50	38	20
55	44	0

At Thebes, Lat. 25° 40′ N., representing Egypt, we find that the amplitude of the sun at rising or setting at the summer solstice will be approximately 26° N. of E. at rising, and 26° N. of W. at setting.

These solstices and their accompaniments are among the striking things in the natural world. At the winter solstice we have the depth of winter, at the summer solstice we have the height of summer; while at the equinoxes we have but transitional changes; in other words, while the solstices point out for us the conditions of greatest heat and greatest cold, the equinoxes point out for us those two times of the year at which the temperature conditions are very nearly equal, although of course in the one case we are saying good-bye to summer and in the other to winter. In Egypt the summer solstice was paramount, for it occurred at the time of the rise of the Nile, the beginning of the Egyptian year.

Did the ancients know anything about these solstices and these equinoxes? Were the almost mythical Hor-shesu or sun-worshippers familiar with the annual course of the sun? That is one of the questions which we have to discuss.

CHAPTER VI.

THE PROBABLE HOR-SHESU WORSHIP.

At the end of the last chapter I referred to the Hor-shesu or followers, that is worshippers, of the Sun-god Horus. I shall have to refer to the traditions relating to them at a later stage, but it is well that I should state here that those personages who preceded the true historic period are considered by De Rougé and others to represent "*le type de l'antiquité la plus reculée.*"

Let us for the moment accept the truth of the various traditions relating to them, and suppose that they left traces of their worship; what, in the light of the last chapter, should we expect to find? The thing most likely to remain would be ancient shrines in all probability serving for the foundation of nobler structures built in later times.

This brings us to the question as to the probabilities of temple-building generally in relation to the heavenly bodies; but before I deal with it, it is important to consider a view first put forward, I believe, by Vitruvius, and repeated by all since his time who have dealt with the question, that the temples were built purely and simply to face the Nile.[1]

The statement is so far from the truth that it is clear that those who have made it had not studied the larger temple-fields. Indeed, we have only to note the conditions at Karnak

[1] "The temples of the gods ought to be so placed that the statue, which has its station in cella, should, if there be nothing to interfere with such a disposition, face the west; in order that those who come to make oblations and offer sacrifices may face the east. . . . When temples are built in the neighbourhood of a river, they should command a view of its banks, like the temples of Egypt upon the borders of the Nile."—*Vitruvius, Civil Architecture, Section I., Chapter V.*

alone to determine whether there is any truth in the view that the temples face the river. We see at once that this idea cannot be true, because we have the chief temples facing in four directions, while the Nile flows only on one side.

Other archæologists who have endeavoured to investigate the orientations of these buildings have found that they practically face in all directions; the statement is that their arrangement is principally characterised by the want of it; they have been put down higgledy-piggledy; there has been a symmetrophobia, mitigated perhaps by a general desire that the temple should face the Nile. This view might be the true one, if stars were not observed as well as the sun.

With regard to all the temples of the ancient world, whether they are located in Egypt or elsewhere, we must never forget that if astronomy is concerned in them at all, we have to deal with the observations of the rising or setting of the heavenly bodies; whereas the modern astronomer cares little for these risings or settings, but deals only with them on the meridian.

The place of rising or setting would be connected with the temple by the direction of the temple's axis.

Now, the directions towards which the temples point are astronomically expressed by their "amplitudes"—that is, the distance in degrees from the east or west point of the horizon. For instance, a temple facing east would have an amplitude of zero from the east point. If we suppose a temple oriented to the north, it would have an amplitude of 90°; if halfway between the east and north, the amplitude would be 45° north of east, and so on. So that it is possible to express the amplitude of a temple in such a way that the temples in the same or different countries or localities, with the same or equivalent amplitudes, may be classified; and the more temples which can be thus brought together, the more likely is any law relating to their structure to come out.

Let us take this, then, as a general principle. Now how would it be carried out?

It becomes pretty obvious, when we consider the conditions of things in these early times, that the stars would be the objects which would first commend themselves to the attention of temple builders, for the reason that the movements and rising- and setting-places of the various planets by night, and of the sun by day, would appear to be so erratic, so long as the order of their movements was not known.

To go a step further. It is clear in the first place that no one would think of orienting a temple to the moon, as there is so little constancy about its path in the sky, and, therefore, in its place of rising or setting. If the temple caught it each month, the intervals between which this occurrence takes place would vary very considerably, and in early times would have been impossible to predict. Similarly it would not be worth while to orient temples to the planets. But when we come to the stars, the thing is different. A few years' observations would have appeared to demonstrate the absolute changelessness of the places of rising and setting of the same stars. It is true that this result would have been found to be erroneous when a long period of time had elapsed and when observation became more accurate; but for hundreds of years the stars would certainly appear to represent fixity, while the movements of sun, moon and planets would seem to be bound by no law.

Before, then, the yearly apparent movements of the sun had been fully made out, observations of a star rising or setting *with the sun* at some critical time of the agricultural cycle, say sowing-time or harvest, would be of the highest importance, and would secure the work being done at the right time of the—to the early peoples—still unformulated year.

If a star was chosen in or near the ecliptic, sooner or later the sun-light as well as the star-light would enter the temple, and the use of a solar temple might have thus been suggested even before the solstices or equinoxes had been thoroughly grasped.

There is no doubt that if we are justified in assuming that the stars were first observed, the next thing that would strike the early astronomers would be the regularity of the annual movement of the sun; the critical times of the sun's movements as related either to their agriculture, or their festivals, or to the year; the equinoxes and the solstices, would soon have revealed themselves to these early observers, if for no other reason than that they were connected in some way or other with some of the important conditions of their environment.

After a certain time, solar temples, if built at all, would be oriented either to the sun at some critical time of the agricultural—or religious—year, or to the solstices and equinoxes. But at first, until the fixity of the sun's yearly movements and especially the solstices and equinoxes had been recognised, it would have seemed as useless to direct a temple to the sun as to the moon. After a time, however, when the solstices and equinoxes had been made out, it would soon have been found that a temple once directed to the sun's rising-place at harvest or sowing time, or at a solstice or an equinox, would continue for a long period to mark those critical points in the sun's yearly course; and when this yearly course had been finally made out it would soon be observed that the sun at any part of the agricultural year was as constant (indeed, as we now know, more constant) in its rising- and setting-place as a star.

But dealing with *sun*-worshippers, and endeavouring to think out what the earliest observers probably would try to do in the case of a *solar* temple, we see that, in all likelihood, they would orient it to observe the sun at one of the chief points in the year which could be best marked. I have said "which could be best marked," but how was this to be done? Evidently, if terrestrial things were to be assisted, the marking must have been by something exterrestrial, otherwise they would have been reasoning in a circle; and moreover we must take for granted that what was wanted was a warning of what was to be done.

Now, in the earliest times, as I have said, the constant movements of the stars would have stood out in strong contrast to the inconstant movements of the sun, and I think that there can be little doubt that the first fixing of any point in the year was by the rising or setting of some star at sunrise—or possibly sunset.

It is obvious that this might have gone on even before the solstices and equinoxes were recognised.

When this came about, then temples might have been directed to the sun at a solstice or an equinox.

Was it difficult to do this? Did it indicate that the people who built such temples were great astronomers? Nothing of the kind; nothing is more easy to determine than a solstice or an equinox.

Let us take the solstice first. We know that at the summer solstice the sun rises and sets furthest to the north, at the winter solstice furthest to the south. We have only from any point to set up a line of stakes before the time of the solstice, and then alter the line of them day by day as the sun gets further to the north or south, until no alteration is wanted. The solstice has been found.

There is another way of doing it. Take a vertical rod. Such a rod, which I may state is sometimes called a *gnomon* and used to measure time, may be used with another object: we may observe the length of the shadow cast by the sun when it is lowest at the winter solstice, and when it is highest; at these two positions of the sun obviously the lengths of the shadows thrown will be different. When the noon-sun is nearest overhead in the summer the length of the shadow will be least, when the sun is most removed from the zenith the shadow will be longest.

The day on which the shortest shadow is thrown at noon will define the summer solstice; when the shadow is longest we shall have the winter solstice.

This, in fact, was the method adopted by the Chinese to determine the solstices, and from it very early they found a value of the obliquity of the ecliptic.

It may be said that this is only a statement, and that the record has been falsified; some years ago anyone who was driven by facts to come to the conclusion that any very considerable antiquity was possible in these observations met with very great difficulty. But the shortest and the longest shadows recorded (1100 years B.C.) do not really represent the true lengths at present. If anyone had forged these observations he would state such lengths as people would find to-day or to-morrow, but the lengths given were different from those which would be found to-day. Laplace, who gave considerable attention to this matter, determined what the real obliquity was at that time, and proved that the record does represent an actual observation, and not one which had been made in later years.[1]

Next suppose an ancient Egyptian wished to determine the time of an equinox. We know from the Egyptian tombs that their stock-in-trade, so far as building went, was very

[1] *See* Biot, "Études sur l'Astronomie Indienne," p. 293.

considerable; they had squares, they had plumb-lines, they had scales, and all that sort of thing, just as we have. He would first of all make a platform quite flat; he could do that by means of the square or plumb-line; then he would get a ruler with pretty sharp edges (and such rulers are found in their tombs), and in the morning of any day he would direct this ruler to the position of the sun when it was rising, and he would from a given point draw a line towards the sun; he would do the same thing in the evening when the sun set; he would bisect the angle made by these two lines, and it would give him naturally a north and south line, and a right angle to this would give him east and west. So that from observations of the sun on any one day in the year he would practically be in a position to determine the points at which the sun would rise and set at the equinox—that is, the true east and west points.

Suppose that the sun is rising, let a rod throw a shadow; mark the position of the shadow; at sunset we again note where the shadow falls. If the sun rises exactly in the east and sets exactly in the west, those two shadows will be continuous, and we shall have made an observation at the absolute equinox. But suppose the sun not at the equinox, a line joining the ends of the shadows equally long before and after noon will be an east and west line.

It is true that there may be a slight error unless we are very careful about the time of the year at which we make the observations, because when the sun is exactly east or west at the time of rising or setting it changes its declination most quickly. So it is better to make the above observations of the sun nearer the solstices than the equinoxes, for the reason stated.[1]

[1] *See* Biot, "Sur divers points d'Astronomie ancienne : Memoires, Académie des Sciences," 1846, p. 47.

We have now got so far. If the Egyptians worshipped the sun and built temples to it, they would be more likely to choose the times of the solstices and the equinoxes than any other after its annual movement had been made out.

Is it possible to bring any tests to bear to see whether they did this or not? Certainly: examine the temples which still remain, and where they have disappeared examine the *temenos* walls which still exist as mounds in many cases.

Suppose we take, to begin with, as before, that region of the earth's surface in the Nile valley with a latitude of about 26° N. The temples will have an amplitude of about 26° N. or S. if they have anything to do with the sun at the solstices. Any structures built to observe the sun will have an east and west aspect true if they have anything to do with the sun at the equinoxes. Dealing with a solstitial temple, the first thing to observe is the amplitude of the temple, which must depend upon the latitude in which it was wished to note the rising or setting of the sun at either of the solstices. If we take the latitude 26° N., which is very nearly the latitude of Thebes, the amplitude has to be 26° as stated above; so that a temple at Thebes having an amplitude of 26° would be very likely to have been oriented to the sun at the moment that it was as far from the equator as it could be—*i.e.*, at the time of the longest day of the year—in which case we should be dealing with the summer or northern solstice; or of the shortest day of the year, if dealing with the winter or southern solstice.

As we deal with higher latitudes, we gradually increase the amplitude, until, if we go as far as the latitude of the North Cape, the sun at the summer solstice, as everybody knows, has no amplitude either at rising or setting, because it passes clear above the horizon altogether, and is seen at midnight.

These are the conditions which will define for us a solstitial solar temple. We see the amplitude of the temple must vary with the latitude of the place where it is erected.

But the temples directed to the sun at an equinox will be directed to an amplitude of 0: that is, they will point E. or W., and this will be the case in all latitudes.

The orientation of a temple directed to the sun at neither the solstices nor the equinoxes will have an amplitude less than the solstitial amplitude at the place.

As a matter of fact, as I shall show in the sequel, some of the temples recognised as temples of the sun in the inscriptions are of this latter class.

CHAPTER VII.

METHODS OF DETERMINING THE ORIENTATION OF TEMPLES.

THIS brings us at once to a practical point. It will be asked, How can such an inquiry be prosecuted? How can the amplitudes of the temples be determined?

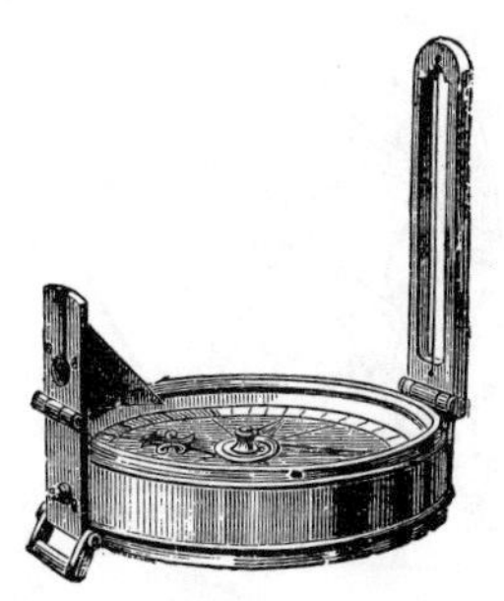

AZIMUTH COMPASS.

Nothing is easier. An azimuth compass is all that is necessary for all but the most accurate inquiries.

The azimuth compass is an instrument familiar to many;

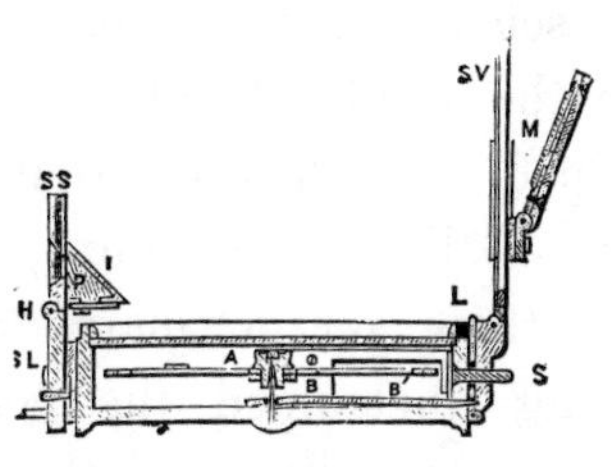

SECTION OF AZIMUTH COMPASS.

A, needle and card; P, prism; S V, directrix or frame carrying a wire directed to the object and seen over the prism while the prism reflects to the eye the division of the scale underneath it.

it consists of a magnetic needle fastened to a card carrying a circle divided into 360°, which can be conveniently read by a prism when the instrument is turned toward any definite direction marked by a vertical wire. Its use depends upon the fact that at the same place and at the same time all magnetic needles point in the same direction, and the variation for the true north and south direction is either supposed to be known or can be found by observation.

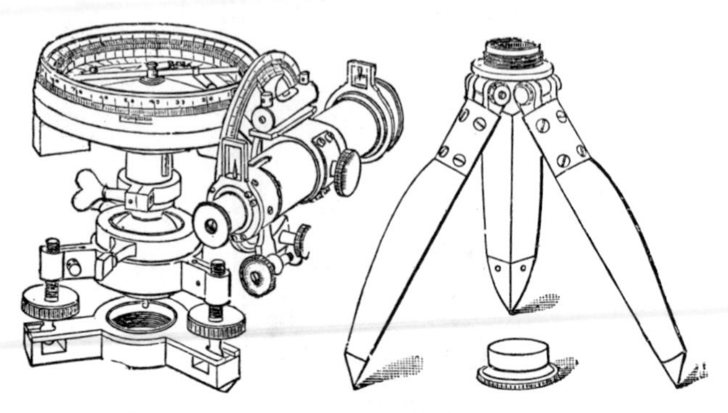

THEODOLITE FOR DETERMINING AZIMUTH AND ALTITUDES.

A theodolite armed with a delicately hung magnetic needle, which can be rotated on a vertical axis, will do still better; it has first of all to be levelled. There is a little telescope with which we can see along the line. When we wish, for instance, to observe the amplitude of a temple, the theodolite is set up on its tripod in such a position that we can look along a temple wall or line of columns, etc., by means of the telescope. We then get a magnetic reading of the direction after having unclamped the compass; this gives the angle made between the line and the magnetic north (or south), as in the azimuth compass.

What we really do by means of such an instrument is to

determine the astronomical meridian by means of a magnetic meridian. Here some definitions will not be out of place.

The *meridian* (*meridies* = midday) of any place is the great circle of the heavens which passes through the zenith (the point overhead) at that place and the poles of the celestial sphere.

The *meridian line* at any place is the intersection of the plane of the meridian with the plane of the horizon at that place, or, in other words, it is the line joining the north and south points. If we have the proper instruments, we can determine the meridian line astronomically at any place by one of the following methods:—

(1) If only an approximate position is required, the best means of determining it is by fixing the direction of the sun or a star when it has the greatest altitude. The instrument to be used for this purpose would be a small theodolite with both a vertical and horizontal circle, and provided also with tangent screws to give slow motion to each of the circles as required.

By using stars of both high and low altitudes, a greater exactness can be obtained, but, after all, the method only gives a first approximation, as its weakness lies in the very slow change of altitude as the meridian is approached.

(2) A much more accurate method is that of observing with an altitude and azimuth instrument the azimuth (*i.e.*, its angular distance east or west of the north or south) of a star when at the same altitude east and west of the meridian. If the mean of the two readings given by the azimuth circle be taken, the resulting reading indicates the direction of the meridian.

If we employ the sun in place of a star, its change of declination during the interval between the observations must be taken into account.

(3) To find the meridian line by means of the pole star is a simple and accurate method, as a value can be obtained at

any time at night by a simple altitude, provided the time of observation is known.[1]

If these means of directly determining the astronomical meridian line are not available, then we have to do it indirectly by using the magnetic meridian in the first instance.

If we take a magnetic needle and balance it horizontally on a vertical pivot, its ends will be directed to two points on the horizon. By drawing a great circle through these two points and the zenith point of the place, we obtain the *magnetic meridian*.

The *magnetic meridian line* is the intersection of the plane of the magnetic meridian with the plane of the horizon. The angle between the astronomical and magnetic meridian lines is called the variation, E. or W. according as the needle points to the W. or E. of true—that is, astronomical—north at any particular place at any particular time. The variation may vary from place to place, and always varies from time to time.

The bearing required has, in the first instance, to be determined by the instruments already referred to in relation to the magnetic meridian.

Having made such an observation, the next thing we have to do is to determine the astronomical or true north, which is the only thing of value.

If the magnetic variation has been determined for the

[1] For a detailed account of the way in which the formula in use has been obtained, the reader had better turn to Vol. I., p. 253, of Chauvenet's "Spherical and Practical Astronomy."

If we denote the latitude by ϕ,

and let p = the star's polar distance,
α = ,, ,, right ascension,
Θ = sidereal time of observation,
h = the star's altitude,
t = ,, ,, hour angle;

then, knowing that

$$t = \Theta - \alpha,$$

the formula may be written as follows:—

$$\phi = h - p \cos. t + \tfrac{1}{2} p^2 \sin. 1'' \sin.^2 t \tan. h.$$

The Nautical Almanac gives tables to facilitate the computations involved, but greater exactness is obtained by direct computation.

region, we may use a map. Such a map as that shown below gives us the lines along which in the British Isles the compass variation west of north reaches certain values. From such a map for Egypt we learn that in 1798 a magnet swung along a line extending from a little to the west of

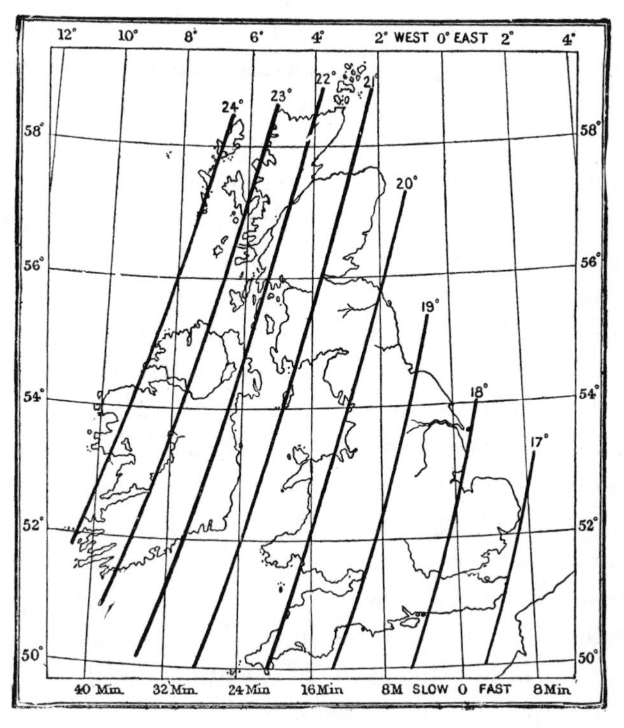

MAGNETIC MAP OF THE BRITISH ISLES, SHOWING THE VARIATION AT DIFFERENT POINTS.

Cairo to the second cataract would have had a variation of $11\frac{1}{2}°$ to the west; in 1844 of $8\frac{1}{2}°$ to the west; and at the present time the variation is such that observations made along the same part of the Nile valley will have a variation closely approximating $4\frac{1}{2}°$ to the west. By means of such a map it is quite possible to get approximately the astronomical bearings of all temples which were observed by the French in 1798 or by the Germans in 1844, or which can be observed in the present day, provided always that there is no local magnetic attraction.

If we are not fortunate enough to possess such a map, the methods previously referred to for obtaining the astronomical north must be employed; observing the direction in which the sun culminates at noon will give us the south point astronomically; from observations of the pole star at night the astronomical north can also be determined. From the former of these observations the magnetic variation is obtained without any difficulty, even in the absence of accurate local time. When this is available other methods are applicable.

It is sad to think how much time is lost in the investigation of a great many of these questions for the reason that the published observations were made only with reference to the magnetic north, which is vastly different at different places, and is always varying. Few indeed have tried to get at the astronomical conditions of the problem. Had this been done with minute accuracy in all cases, either by the French or Prussian Commissions to which I have referred, it is perfectly certain that the solstitial orientation of Karnak and other temples, which I shall have to mention, would have been long ago known to all scholars.

CHAPTER VIII.

THE EARLIEST SOLAR SHRINES IN EGYPT.

NOT only can an inquiry like that referred to in the previous chapter be prosecuted—*it has been prosecuted.*

The French and Prussian Governments have vied with each other in the honourable rivalry of mapping and describing the monuments. The French went to Egypt at the end of the last century, while the Scientific Commission which accompanied the army, a Commission appointed by the Institute of France, published a series of volumes containing plans of all the chief temples in the valley of the Nile as far south as Philæ.

In the year 1844, some time after Champollion had led the way in deciphering the hieroglyphics, we became almost equally indebted to the Prussian Government, who also sent out a Commission to Egypt, under Lepsius, which equalled the French one in the importance of the results of the explorations; in the care with which the observations were made, and in the perfection with which they were recorded. In attempting to get information from ancient temples on the points to which I have referred, there is, therefore, a large amount of information available; and it is wise to study the region round and below Thebes where the information is so abundant and is ready to our hand.

First, then, with regard to the existence of solar temples. Dealing with the monumental evidence, the answer is absolutely overwhelming. The evidence I bring forward consists of that afforded by some of the very oldest temples that we know of in Egypt. Among the most ancient and sacred fanes was one at Annu, On, or Heliopolis, which, the tradition runs, was founded by the Shesu-Hor before the time of Mena; Mena, as we have seen, having reigned at a date certainly not less than 4000, and possibly 5000 years B.C.

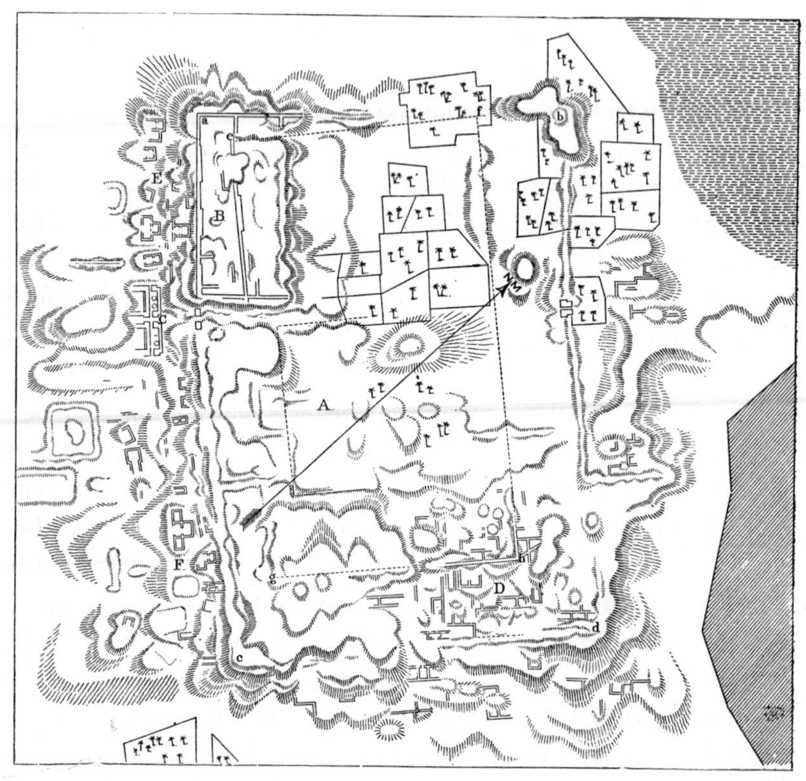

PLAN OF THE MOUNDS AT ABYDOS. (*From Mariette.*)

The Nile valley holds other solar temples besides that we have named at Heliopolis. Abydos was another of the holiest places in Egypt in the very earliest times.

Since the temples and temple mounds at Abydos can be

better made out than those at Heliopolis, I will take them first. The orientations given by different authors are so conflicting that no certainty can be claimed, but it is possible that at Abydos one of the mounds is not far from the amplitude shown in the tables for the sun in the Nile valley at sunset at the summer solstice. If this were so, the Egyptians who were employed in building the temple must have known exactly what they were going to do.

At Heliopolis, as I have hinted, the matter is still less certain. Almost every trace of the temple has disappeared, but of remains of temenos walls in 1844, when the site was studied by Lepsius, there were plenty. At Karnak, where both temples and temenos walls remain, we can see how closely the walls reflect the orientation of the included temples, even when they seem most liable to the suggestion of symmetrophobia. I have before stated that the Egyptians have been accused of hating every regular figure, and the irregular figures at Karnak are very remarkable; in the boundary walls of the temple of Amen-Rā there are two obtuse angles; round the Mut temple we also have walls, and there again this hatred of similarity seems to come out, for we have one obtuse and one acute angle. But if we examine the thing a little carefully, we find that there is a good deal of method in this apparent irregularity. The wall of the temple of Amen-Rā is parallel to the face of the temple or at right angles to its length. One wall of Mut is perfectly parallel to the face of the temple or at right angles to the sphinxes. And the reason that we do not get right angles at one end of the wall is that the walls of the temple at Mut are parallel to the chief wall of the temple of Amen-Rā. Surely it must be that, before these walls were built, it was understood that there was a combined worship; that they stood or fell together. One thing was not attempted in one temple and another thing in another, but the worship of each was reflected in the other. If this be true, there was no hatred of symmetry, but a definite and admirable reason why these walls should be built as they were.

With the knowledge we possess of both temenos walls and temples at Karnak, and of the, I may almost say, symbolism of

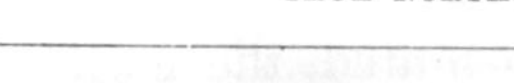

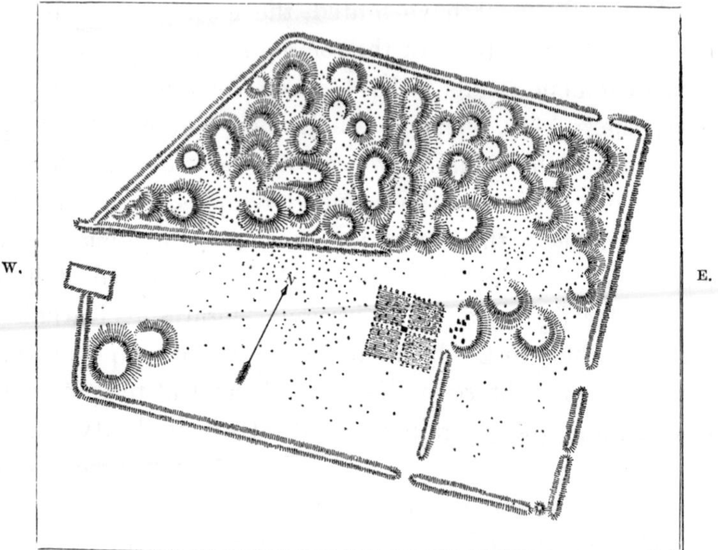

THE MOUNDS AND OBELISK AT ANNU.

the former, it is fair to conclude that when temples have gone we may yet get help from the walls. The walls at Heliopolis are the most extraordinary I have met with in Egypt, as may be gathered from the accompanying reduction of Lepsius' map.

The arrow in Lepsius' plan is so wrongly placed that the plan is very misleading. It follows from Captain Lyons' observations and my own that the longest mound heads 14° N. of W.

to 14° S. of E. within a degree; the condition of the mounds renders more accurate measures impossible.[1]

It is to be gathered from the inscriptions that the temple within these mounds, now only represented by its solitary obelisk, was styled a sanctuary or temple of the sun.[2]

As the orientation of the N. and S. faces of the obelisk is 13° N. of W., the sun's declination must have been 11° N. The times of our year marked by it, therefore, were 18th April and 24th August. But it must not be forgotten that the temple may have been built originally to watch the rising or setting of a star which occupied the declination named, and possibly, though not necessarily, at some other time of the year. I shall return to this subject.

If Maspero and the great authorities in Egyptology are right—namely, that the Annu temple was founded before 4000 B.C.—the above figures drive us to the conclusion that we have

[1] Since I left Egypt, in February, 1893, Captain Lyons has been good enough to comply with my request to repeat the observations. I give the following extract from his letter :—

"The mounds are only within a degree, as it is only the general direction which can be taken. South mound old temenos wall, 289½° mag. bearing = 19½° N. of W.
Wall at right angles ... 189° mag. bearing = 71° S. of W.

Going to the West mound there are two higher humps with an opening between them, tons of limestone chips, sandstone blocks with Rameses II.'s name; so that I take this for the site of the great pylon. It is exactly opposite the obelisk, and distant, I should guess, 600 yards. Site of S. pylon to obelisk, 106½° mag. bearing = 16½° S. of E.
Pole of N. pylon to obelisk, 109½° mag. bearing = 19½° S. of E.

So I think probably the remaining obelisk is the northern one (*cf.* Horner, "Phil. Trans.," MDCCCLV., pp. 124 and 131), and the temple axis was directed 289½° mag. bearing with corr. 5½° = 284° = 14° N. of West true amplitude."

[2] Amenemāt I., the founder of the sanctuary of the sun, entreats, after he has begun the great work (which was not finished till the time of his son, Usertesen), "May it not perish by the vicissitudes of time, may that which is made endure!" This desire of a great king which has come down to us through the leathern roll now preserved at Berlin, has not been fulfilled; for of his magnificent structure, built for all eternity, nothing remains but the obelisk we have seen, and a few blocks of stone scarcely worth mentioning. The Persian Cambyses is unjustly accused of having destroyed the temple and city of the sun, for the city was minutely described in detail long after his time, and the temple was still flourishing; nay, many remains of the sanctuary, that have now long since vanished, were described even by Arab authors.—Ebers, "Egypt," p. 190.

in this temple a building which was orientated to the sun, *not* at a solstice, some 6000 years ago.

So much for two of the places known to be of the highest antiquity in Egypt. There remains another locality supposed to date from more modern times—I refer to Thebes. It is here that evidence of the most certain kind with regard to the solstitial temples is to be found.

At Karnak itself there are several temples so oriented, chief among them the magnificent Temple of Amen-Rā, one of the wonders of the world, to which a special chapter must be devoted. Suffice it to say here that the amplitude of the point to which the axis of the great temple of Amen-Rā points is 26° N. of W., which we learn from the table already given is the amplitude of the place of sunset at the summer solstice in the latitude of Thebes. The amplitude of the point to which the axis of an attached small temple points is 26° S. of E., exactly the position of sunrise at the winter solstice.

It must not be forgotten in this connection that the Colossi of the plain on the other side of the river, and the associated temple, also face the place of sunrise at the winter solstice.

The list of solar solstitial temples, so far probably traced, is as follows :—

Place and Temple.		Amplitude.	Declination.	Date.
S.E. Temples.	Kasr Kerun	27° S. of E.	S. $23\frac{1}{4}$°	
	Karnak (O)	$26\frac{1}{2}$° S. of E.	S. $23\frac{3}{4}$°	
	Memnonia (Avenue of Sphinxes) (orientation not to $\frac{1}{2}$°)	$27\frac{1}{2}$° S. of E.	S. $24\frac{1}{2}$°	
S.W. Temples.	Erment	$27\frac{1}{2}$° S. of W.	S. $24\frac{1}{2}$°	
N.W. Temples.	Karnak (Q. K)	$26\frac{1}{2}$° N. of W.	N. $23\frac{1}{2}$°	
	Karnak (U)	$27\frac{1}{2}$° N. of W.	N. $24\frac{1}{2}$°	

THE COLOSSI OF THE PLAIN AT THEBES AT HIGH NILE, ORIENTED TO THE SUNRISE AT THE WINTER SOLSTICE.

(These are statues of Amen-hetep III., and are monoliths 60 feet high.)

We have seen that it did not require any great amount of astronomical knowledge to determine either the moment of the solstice or the moment of the equinox. The most natural thing

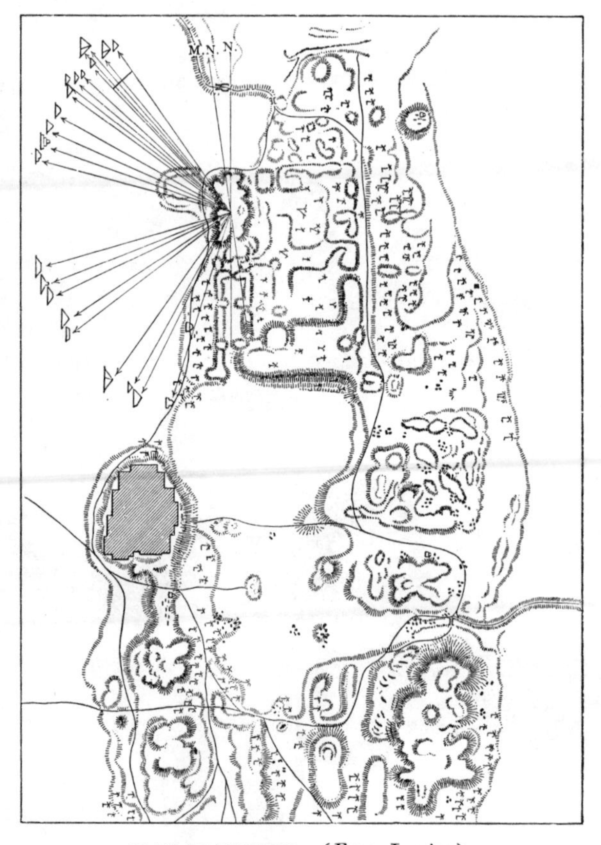

PLAN OF MEMPHIS. (*From Lepsius.*)

to begin with was the observation of the solstice, for the reason that at the solstice the sun can be watched day after day getting more and more north or more and more south until it comes to a standstill. But for the observation of the equinox, of course, the sun is moving most rapidly either north or south, and therefore it would be more difficult to determine in those days the exact moment.

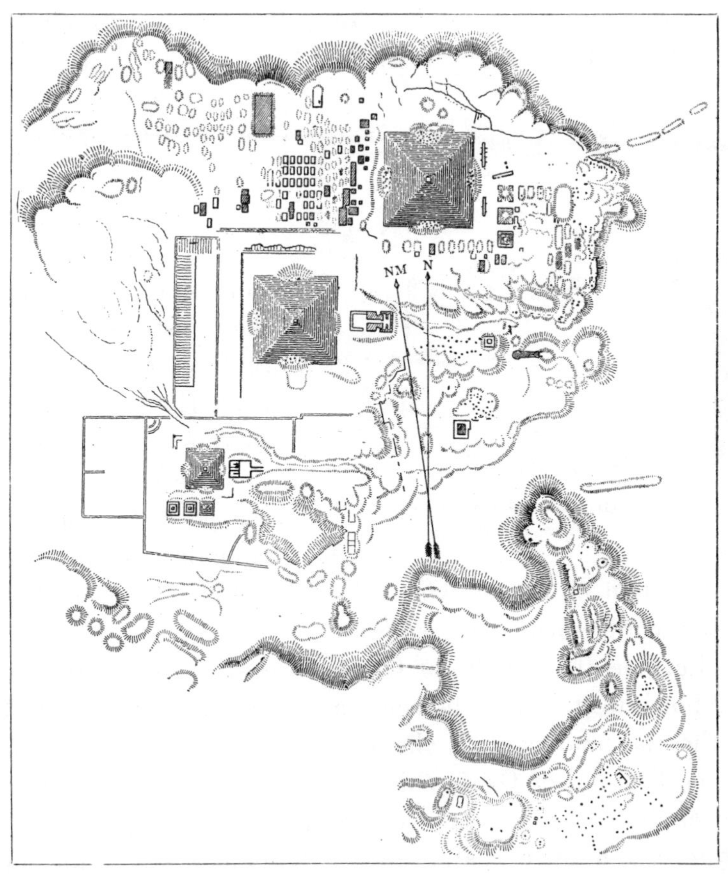

EAST AND WEST PYRAMIDS AND TEMPLES AT GÎZEH. (*From Lepsius.*)

We next come to the question as to whether any buildings were erected from an equinoctial point of view—that is, buildings oriented east and west.

Nothing is more remarkable than to go from the description

and the plans of such temples as we have seen at Abydos, Annu, and Karnak, to regions where, apparently, the thought is totally and completely different, such as we find on the Pyramid Plains at Gîzeh, at Memphis, Tanis, Saïs, and Bubastis. The orientation lines of the German surveyors show beyond all question that the pyramids and some of the temenos walls at the places named are just as true to the sun-rising at the equinoxes as the temples referred to at Karnak were to the sun-rising and setting at the solstices, and the Sphinx was merely a mysterious nondescript sort of thing which was there watching for the rising of the sun at an equinox, as the Colossi of the plain at Thebes were watching for the rising of the sun at the winter solstice.

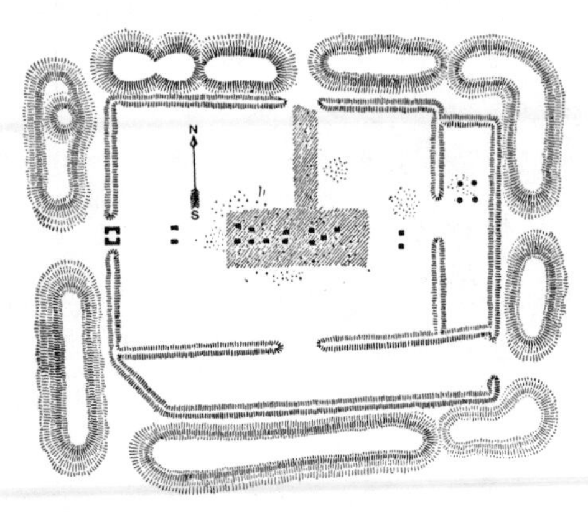

TEMPLE AND TEMENOS WALLS OF TANIS.
(*From Lepsius.*)

Further, the temples at Gîzeh, instead of being oriented to the north-west and to the south-east, are just as truly oriented to the east and west as the Pyramids themselves. We have either Temples of Osiris pointing to the sunset at the equinox, or temples of Isis pointing to the sunrise at the equinox, but in either case built in relation to the Pyramids. As an indication of the importance of the considerations with which we are now dealing, I may mention that it is suggested by them that the building near the Sphinx is really a crypt of a temple of Isis or Osiris. This is a view which may change the ideas generally held with regard to its age to the extent of something like a thousand years. It has been imagined that it was at least one thousand years older than the second Pyramid; but if it be ultimately proved that this is really a temple of Isis or Osiris, then since it was built in just as strict relation to the side

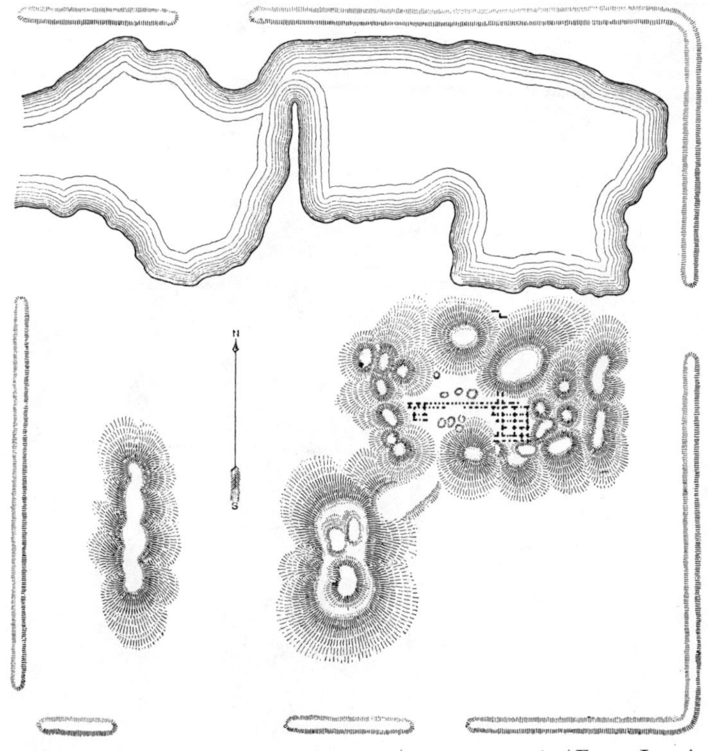

TEMPLE AND TEMENOS WALLS OF SAIS (SA-EL-HAGAR). (*From Lepsius.*)

of the Pyramid as the temple near the Pyramid was to its centre, both temples were most probably built at the same time as the Pyramid itself. However this may be, the important thing is that when we pass from Thebes, and possibly Abydos, to the Pyramids at Memphis, to Saïs and Tanis, we find a

THE TEMPLE NEAR THE SPHINX, LOOKING WEST (TRUE), SHOWING ITS RELATION TO THE SOUTH FACE OF THE SECOND PYRAMID.
(From a photograph by Mr. Fearing.)

solstitial orientation changed to an equinoctial one. *There is a fundamental change of astronomical thought.*

I confess I am impressed by this distinction; from the astronomical point of view it is so fundamental that almost a difference of race is required to explain it. I say this advisedly, although I know creed can go a great way, because among these early peoples their astronomy was chiefly a means to an end. It was not a story of abstract conceptions, or the mere expression of interesting facts whether used for religious purposes or not. The end was a calendar, of festivals and holydays if you will, but a calendar which would allow their tillage and harvest to prosper.

Now, it is almost impossible to suppose that those who worshipped the sun at the solstice did not begin the year at the solstice. It is, of course, equally difficult to believe that those who preferred to range themselves as equinoctials did not begin the year at an equinox. Both these practices could hardly go on in the case of the same race in the same country, least of all in the valley where an annual inundation marked the solstice.

I shall show subsequently how the rise of the Nile, which took place at the summer solstice, not only dominated the industry, but the astronomy and religion of Egypt; and I was much interested in hearing from my friend Dr. Wallis Budge that the rise of the Tigris and Euphrates takes place not far from the spring equinox. This may have dominated the Babylonian calendar as effectually as the date of the Nile-rise dominated the Egyptian. If so, we have a valuable hint as to the origin of the equinoctial cult at Gîzeh and elsewhere, which in all probability was interpolated after the non-equinoctial worship had been first founded at Annu, Abydos, and possibly Thebes.

CHAPTER IX.

OTHER SIMILAR SHRINES ELSEWHERE.

THE observations which have been made in Babylonia are very discordant among themselves, and at present it is impossible to say, from the monuments in any part of the region along the Tigris valley, whether the temples indicate that the solstices were familiar to the Babylonians.

The ancient cities which have so far been excavated and the modern names of the sites are as follows:

Nineveh	=	Kouyunjik.
Babylon	=	Birs Nimrûd.
Calah	=	Nimrûd.
Erech	=	Warka.
Ur of the Chaldees	=	Mukeyyer.
Ashur	=	Kalat Sherkât.
Dur Sarginu	=	Khorsabad.

Let us take, for instance, the region in the valley near where the Upper Zab joins the main stream. We gather from the map published in 1867 by Place,[1] that Nimrûd, the modern Calah, is near the junction, while the mounds of Kouyunjik, Môsul, and Khorsabad, representing the ancient Dur Sarginu, are to the north (36° N. latitude). There are two other mounds shown on the map at Djigan and Tel Hakoab.

Now, by inspection it is quite clear that none of the mounds except that of Nimrûd lie east and west. It becomes important, therefore, to determine their orientation; but, alas! this is

[1] "Ninive et l'Assyrie," par Victor Place. Imprimerie Imperiale, 1867.

nearly impossible with the sole exception of Khorsabad, for no measures appear to have been made.

At first sight the matter seems more hopeful in the case of Khorsabad, for we have not only the plans of Place, but those of Botta and Flandin.[1] The plans seem oriented with care, so far as the existence of a compass direction is concerned—for that is present while it too often is lacking in such productions—but in neither series is it stated whether N. means true or magnetic north.

Both observers noted a well-marked temple facing N.E., and also an "observatory." About the temple there can be no mistake, for the fair-way of the light to it is carefully preserved, and there is a flight of wide steps on the north-east side of it.

Place gives the orientation 37° N. of E. in one plan and 39° in another. Botta and Flandin give 31½° in one plan and 32° in another! Now, the change in the magnetic variation between 1849 and 1867 will not explain this difference, nor indeed can it be accounted for by supposing that the magnetic north is in question in one set of plans and the true north in the other;[2] and it is clear that no perfectly certain conclusion can be arrived at till this work has been done over again. But it is known that M. Flandin was a skilled surveyor, and we have the remarkable fact, that if we take his value, *we have the amplitude of the sun at the summer solstice in the latitude of Nineveh !*

[1] "Monument de Ninive," par Botta and Flandin. Imprimerie Nationale, 1849.

[2] From a magnetic chart which has been prepared for me by the kindness of Captain Creak, R.N., F.R.S., of the Hydrographic Department of the Admiralty, it seems that the variation at Nineveh and Babylon may be taken as follows :—

	Nineveh.	Babylon.
1800 ...	8° 25′ W.	8° 25′ W.
1900 ...	0°	0° 25 W.

The values for intermediate dates may be roughly arrived at by an interpolation curve

I certainly think the temple may be accepted as a solstitial solar temple provisionally; and if so, the question is raised whether the structures in Assyria, supposed to be oriented so that the *angles* face the cardinal points, are not all of them oriented to the sun at a solstice or to some other heavenly body. Certainly we must have more definite measures before the statement generally made can be accepted as final.

When we leave Assyria we find other countries, it is true still farther afield, in which the existence of solstitial temples of a great antiquity of foundation is fully recognised.

The great temple of the sun at Pekin is oriented to the winter solstice. The ceremonials which take place there are thus described by Edkins:—

"The most important of all the State observances of China is the sacrifice at the winter solstice, performed in the open air at the south altar of the Temple of Heaven, December 21st. The altar is called Nan-Tan, 'south mound,' or Yuenkieu, 'round hillock'—both names of the greatest antiquity.

"Here also are offered prayers for rain in the early summer. The altar is a beautiful marble structure, ascended by twenty-seven steps, and ornamented by circular balustrades on each of its three terraces. There is another on the north side of somewhat smaller dimensions, called the Ch'i-ku-t'an, or altar for prayer on behalf of grain. On it is raised a magnificent triple-roofed circular structure 99 feet in height, which constitutes the most conspicuous object in the *tout ensemble*, and is that which is called by foreigners the Temple of Heaven. It is the hall of prayer for a propitious year, and here, early in the spring, the prayer and sacrifice for that object are prosecuted. These structures are deeply enshrined in a thick cypress grove, reminding the visitor of the custom which formerly prevailed among the heathen nations of the Old Testament, and of the solemn shade which surrounded some celebrated temples of ancient Greece."

The Temple of Heaven is thus described:—

"The south altar, the most important of all Chinese religious structures, has the following dimensions: It consists of a triple circular terrace, 210 feet wide at the base, 150 in the middle, and 90 at the top. In these, notice the multiples of three: $3 \times 3 = 9$, $3 \times 5 = 15$, $3 \times 7 = 21$. The heights of the three terraces, upper, middle, and lower, are 5·72 feet,

6·23 feet, and 5 feet respectively. At the times of sacrificing, the tablets to heaven and to the Emperor's ancestors are placed on the top; they are 2 feet 5 inches long, and 5 inches wide. The title is in gilt letters; that of heaven faces the south, and those of the ancestors east and west. The Emperor, with his immediate suite, kneels in front of the tablet of Shang-Ti and faces the north. The platform is laid with marble stones, forming nine concentric circles; the inner circle consists of nine stones, cut so as to fit with close edges round the central stone, which is a perfect circle. Here the Emperor kneels, and is surrounded first by the circles of the terraces and their enclosing walls, and then by the circle of the horizon. He thus seems to himself and his court to be in the centre of the universe, and turning to the north, assuming the attitude of a subject, he acknowledges in prayer and by his position that he is inferior to heaven, and to heaven alone. Round him on the pavement are the nine circles of as many heavens, consisting of nine stones, then eighteen, then twenty-seven, and so on in successive multiples of nine till the square of nine, the favourite number of Chinese philosophy, is reached in the outermost circle of eighty-one stones.

"The same symbolism is carried throughout the balustrades, the steps, and the two lower terraces of the altar. Four flights of steps of nine each lead down to the middle terrace, where are placed the tablets to the spirits of the sun, moon, and stars and the year god, Tai-sui. The sun and stars take the east, and the moon and Tai-sui the west: the stars are the twenty-eight constellations of the Chinese zodiac, borrowed by the Hindoos soon after the Christian era, and called by them the Naksha-tras; the Tai-sui is a deification of the sixty-year cycle."[1]

We find, then, that the most important temple in China is oriented to the winter solstice.

To mention another instance. It has long been known that Stonehenge is oriented to the rising of the sun at the summer solstice. Its amplitude instead of being 26° is 40° N. of E.; with a latitude of 51°, the 26° azimuth of Thebes is represented by an amplitude of 40° at Stonehenge.

The structure consists of a double circle of stones, with a sort of naos composed of large stones facing a so-called avenue, which is a sunken way between two parallel banks.

[1] "Journeys in North China," Williamson. Vol. II., chap. xvi., by Edkins, p. 253.

This avenue stretches away from the naos in the direction of the solstitial sunrise.

But this is not all. In the avenue, but not in the centre of its width, there is a stone called the "Friar's Heel," so located in relation to the horizon that, according to Mr. Flinders Petrie,[1] who has made careful measurements of the whole structure, it aligned the coming sunrise from a point behind the naos or trilithon. The horizon is invisible at the entrance of the circle, the peak of the heel rising far above it; from behind the circles the peak is below the horizon. Now, from considerations which I shall state at length further on, Mr. Petrie concludes that Stonehenge existed 2000 B C. It must not be forgotten that structures more or less similar to Stonehenge are found along a line from the east on both sides of the Mediterranean.[2]

It will be seen that the use of the marking stone to indicate the direction in which the sun will rise answers exactly the

STONEHENGE, FROM THE NORTH.

[1] "Stonehenge: Plans, Descriptions, and Theories," 1880, p. 20.
[2] Ferguson: "Rude Stone Monuments."

same purpose as the long avenue of majestic columns and pylons in the Egyptian temples. In both cases we had a means of determining the commencement and the succession of years.

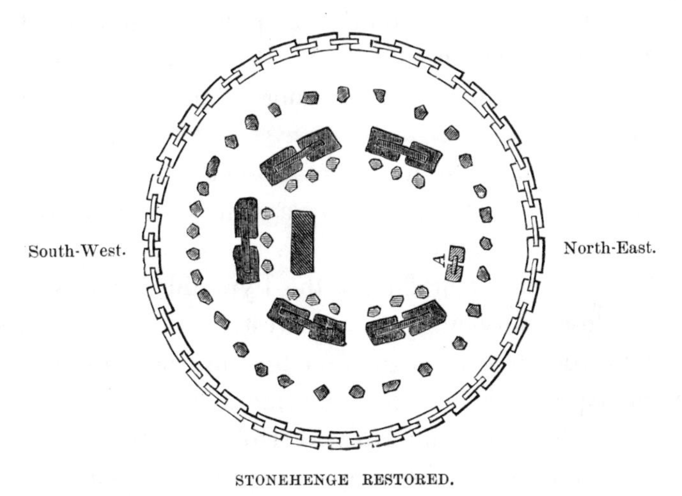

STONEHENGE RESTORED.

Hence, just as surely as the temple of Karnak once pointed to the sun *setting* at the summer solstice, the temple at Stonehenge pointed nearly to the sun *rising* at the summer solstice. Stonehenge, there is little doubt, was so constructed that at sunrise at the same solstice the shadow of one stone fell exactly on the stone in the centre; that observation indicated to the priests that the New Year had begun, and possibly also fires were lighted to flash the news through the country. And in this way it is possible that we have the ultimate origin of the midsummer fires, which have been referred to by so many authors.[1]

We have thus considered solstitial temples scattered widely over the earth's surface far from the Nile Valley.

[1] *See* especially "The Golden Bough," by J. G. Fraser, for the midsummer and Beltaine fires.

We may now return to the equinoctial temples which can still be traced to the N.E. of that valley—the chief ones being those, remains of which still exist at Jerusalem, Baalbek, and Palmyra, where stone was available for the temple builders. These temples were apparently as perfectly squared to the equinox as the Pyramids at Gîzeh. I will take the temple of Jerusalem first, as its history is more complete than that of the others.

We learn from the works of Josephus that as early as Solomon's time the temple at Jerusalem was oriented to the east with care;[1] in other words, the temple at Jerusalem was parallel to the temple of Isis at the Pyramids; it was open to the east, closed absolutely to the west. In plan, as we shall see, it was very like an Egyptian temple, the light from the sun at the equinox being free to come along an open passage, and to get at last into the Holy of Holies. We find that the direction of the axis of the temple shows the existence of a cult connected with the possibility of seeing the sun rise at either the spring or the autumn equinox.

All the doors being opened, the sunlight would penetrate over the high altar, where the sacrifices were offered, into the very Holy of Holies, which we may remember was only entered by the high priest once a year; it could have done that twice a year, but as a matter of fact it was only utilised once; whereas at Karnak the priest would only go into the Holy of Holies once a year, because it was only once illuminated by the sun in each year.

There is evidence, too, that the entrance of the sunlight on the morning of the spring equinox formed part of the ceremonial. The priest being in the naos, the worshippers outside, with their backs to the sun, could see the high priest by means

[1] "Antiquities," b. 8, c. 4, p. 401, Whiston's edition.

of the sunlight reflected from the jewels[1] in his garments, thus referred to by Josephus :—

> "I will now treat of what I before omitted—the garment of the high priest, for he [Moses] left no room for the evil practices of [false] prophets; but if some of that sort should attempt to abuse the Divine authority, he left it to God to be present at His sacrifices when He pleased, and when He pleased to be absent. And he was willing this should be known, not to the Hebrews only, but to those foreigners also who were there. For as to those stones, which we told you before, the high priest bare on his shoulders, which were sardonyxes (and I think it needless to describe their feature, they being known to everybody), *the one of them shined out when God was present at their sacrifices.*[2] I mean that which was of the nature of a button on his right shoulder, bright rays darting out thence, and being seen even by those who were most remote; which splendour yet was not before natural to the stone."

Josephus[3] states that the miraculous shining of the jewels ceased two hundred years before his time, "God having been displeased at the transgression of His laws."

This remark of Josephus quite justifies the assumption that the effect of sunlight on the priest's jewels formed part of the ceremonial, and in this way. In the earliest times there is no doubt that the equinoctial temples were solar temples pure and simple, and the rising sun would always, in fine weather, shine into them at the equinox, which, while they were used as solar temples, marked New Year's Day. The influence of the later Babylonian astronomy, however, at length replaced the sun by the moon, and the year would commence, not at the equinox, but by a new or a full moon near the equinox. If either of these happened *at* the equinox, well and good; but if not, then the sun's declination might be widely different from 0°—it might amount roughly to 10° either N. or S.—and under these circumstances, as the amplitude

[1] Josepnus, "Antiquities" III., c. 8, § 9. [2] The italics are mine—J. N. L.

[3] "Antiquities" III., c. 8, § 9.

would be greater, the sun's light could not enter the temple at all at the date of the feast. More than this, a mistake of a month might be made, or a question of old style and new style might come in, and that of course would make matters worse. In this way, then, the withdrawal of the sunlight from the temple at Jerusalem admits of being astronomically explained.

It seems highly probable that the temple in question was built on a Phenician foundation, for some of the stones exceed 38 feet in length and weigh 90 tons.[1] This remark is suggested by the fact that at Baalbek or Heliopolis, to which I next direct attention, the most ancient and most massive part of the structure is, in all probability, of Phenician origin. To give an idea of its massiveness, which is almost more than Egyptian, it may be stated that there are three stones each about 64 feet long, 13 feet high, and 13 feet thick. There are smaller stones used in the filling in, of the same height and thickness, and 30 feet long.[2] These form the western wall of the original naos or of its support.

Here the orientation is due E.[3] When we come to Palmyra, we find also another temple to the equinoctial sun; but here the sunset, and not the sunrise, is in question —the temple faces due west.

In the whole problem, then, of orientation as I have had to present it, and as it now stands, we seem for the moment to be face to face with two very remarkable and strange things; so strange that the argument may appear far-fetched and worthless, since we are landed in a region apparently very far

[1] Warren : "Underground Jerusalem."

[2] Acosta, in his "History of Indies," lib. vi., p. 459, quoted by Maurice ("Observations Connected with Astronomy and Ancient History and Ruins of Babylon"), states that some of the stones in the Mexican temples to sun and moon measure 38 feet by 18 feet by 6 feet.

[3] *See* "Palmyra and Baalbek." R. Wood, 1827. Plates.

removed from our modern habits of thought. But is this really so? I assume the personification or the deification of the sun: I shall subsequently have to include the stars; I indicate special orientations of buildings devoted to the worship of the sun at one time of the year or another. But really both these things, though they seem improbable, have been carried down to our own day, quite independently of any question relating to Egypt. There is nothing new about them at all, and there is nothing really strange. When we go into an observatory we think nothing of turning our telescope towards Venus, or Jupiter, or Mars. Here we have the deification of the planets. It is perfectly true that this religious treatment of the planets is not of our own day: we have inherited it from the Greeks through the Latins; but we do not think it at all extraordinary that a planet should be called Venus or Jupiter. Thus we of to-day are completely in touch with the old Egyptians, except that the Egyptians were wiser in their generation, and looked after the sun at fixed points in the year and the constant stars instead of the variable planets.

Then, again, take the question of orientation. This is, after all, one which survives among ourselves. All our churches are more or less oriented, which is a remnant of old sun-worship.[1] Any church that is properly built to-day will have its axis

[1] On this point I gather the following information from the article "Orientation" in the "Grand Dictionnaire Universel du 19 Siècle," by M. Pierre Larousse:—"From the fifth century to the time of the Renaissance, the orientation of churches was generally carried out. The mystical reasons furnished by the sacred writers—according to St. John of Damascus and Cassiodorus—were that Jesus on the Cross had His face turned towards the West, hence Christians during prayer must turn to the East to see it. Further, in the sacred writings Jesus is called the East (*Oriens ex alto*). Again, Christians hope to see Christ descending in the East on the last day. Finally, the faithful when turning to the East during prayer establish a difference between themselves and the Jews and heretics, for the Jews when praying turn West, and certain heretics South, and others North, hence the heathen said they were sun-worshippers." In the ninth century there was a strong protest against orientation. Catholic churches were built any way, and it was said, "*Nunc oremus ad omnem partem quia Deus ubique est.*"

pointing to the rising of the sun on the Saint's Day, *i.e.*, a church dedicated to St. John ought not to be parallel to a church dedicated to St. Peter. It is true that there are sometimes local conditions which prevent this; but if the architect knows his business properly he is unhappy unless he can carry

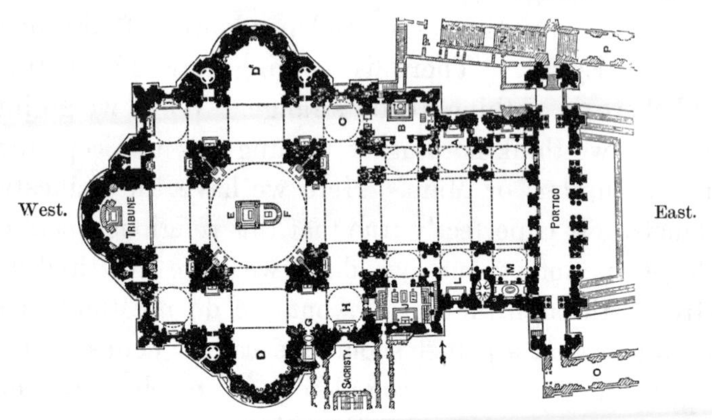

PLAN OF ST. PETER'S AT ROME, SHOWING THE DOOR FACING THE SUNRISE.

out this old-world tradition. But it may be suggested that in our churches the door is always to the west and the altar is always to the east. That is perfectly true, but it is a modern practice. Certainly in the early centuries the churches were all oriented to the sun, so that the light fell on the altar through the eastern doors at sunrise. The late Gilbert Scott, in his "Essay on Church Architecture," gives a very detailed account of these early churches, which in this respect exactly resembled the Egyptian temples.

In regard to old St. Peter's at Rome,[1] we read that "so exactly due east and west was the Basilica that, on the vernal equinox, the great doors of the porch of the quadriporticus were thrown open at sunrise, and also the eastern doors of the church itself, and as the sun rose, its rays passed through the outer

[1] See *Builder*, Jan. 2, 1892.

ST. PETER'S AT ROME; FAÇADE FACING THE EAST (TRUE).

doors, then through the inner doors, and, penetrating straight through the nave, illuminated the High Altar." The present church fulfils the same conditions.

But we have between our own churches and the Egyptian temples a link in the chain which has just been magnificently completed by Mr. Penrose by his study of the Greek temples. These interesting results will occupy us in a later chapter.

CHAPTER X.

THE SOLAR TEMPLE OF AMEN-RĀ AT KARNAK.

So much having been premised concerning the early temple-worship of the sun in Egypt and the adjacent countries, and the survival of some of the ideas connected with it down to our own day, I next propose to describe the finest Egyptian solar temple which remains open to our examination—that of Amen-Rā at Karnak.

Of the chief solar temples referred to in a previous chapter, two have passed away; even the orientation of the one at Heliopolis I was only able to determine by the mounds, assuming them to bear the same relation to the temple as other mounds do, and the remaining obelisk.

The temple at Abydos is also a mound; but in the case of the temple of Amen-Rā at Thebes the case is different: instead of being a mere heap, the orientation of which is obtainable only by the general lie of the remains, this temple is still in such preservation that Lepsius in the year 1844 could give us a large number of details about it, and locate the position of the innumerable courts. Its orientation to the solstice we can claim, as I hope to be able to show, as an early astronomical observation. So it is quite fair to say that, many thousand years ago at all events, the Egyptians were perfectly familiar with the solstices, and therefore more or less fully with the yearly path of the sun.

This temple of Amen-Rā is beyond all question the most majestic ruin in the world. There is a sort of stone avenue in the centre, giving a view towards the north-west, and this axis is something like five hundred yards in length. The whole object of the builder of the great temple at Karnak—one of the most soul-stirring temples which have ever been conceived or built by man—was to preserve that axis absolutely open; and all the wonderful halls of columns and the like, as seen on one side or other of the axis, are merely details; the point being that the axis should be absolutely open, straight, and true. The axis was directed towards the hills on the west side of the Nile, in which are the tombs of the kings. From the external pylon the South-eastern outlook through the ruins shows the whole length of

AXIS OF THE TEMPLE OF AMEN-RĀ FROM THE WESTERN PYLON, LOOKING SOUTH-EAST.

the temple, and we see at the very extremity of the central line a gateway nearly six hundred yards away. This belonged

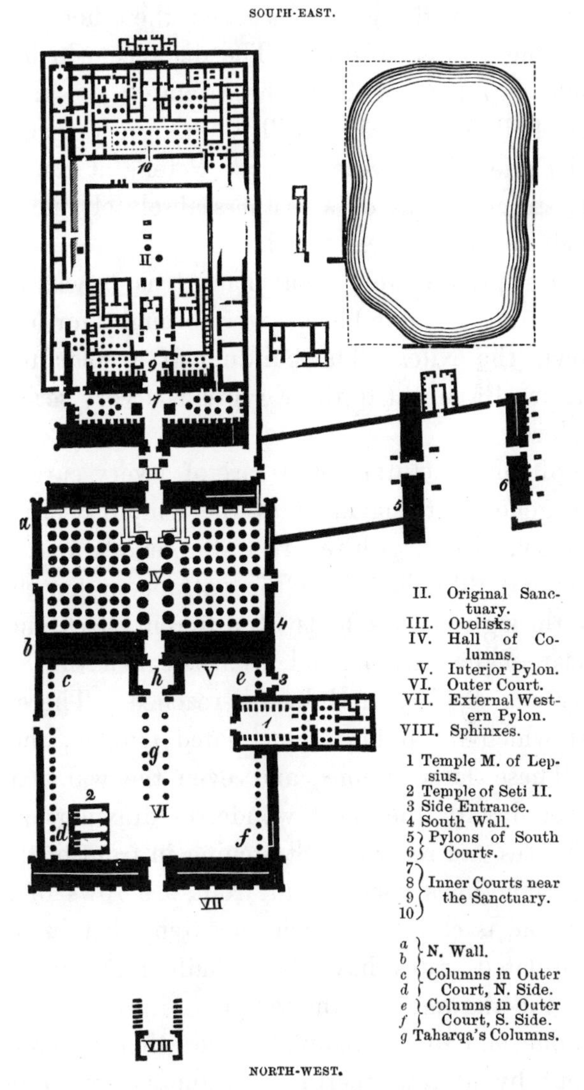

PLAN OF THE TEMPLE OF AMEN-RĀ AND SOME OF ITS SURROUNDINGS, INCLUDING THE SACRED LAKE.

to a temple pointing towards the south-east. There were really two temples in the same line back to back, the chief one facing the sunset at the summer solstice, the other probably the sunrise at the winter solstice. The distance which separates the outside entrances of both these temples is greater than that from Pall Mall to Piccadilly; the great temple covers about twice the area covered by St. Peter's at Rome, so that the whole structure was of a vastness absolutely unapproached in the modern ecclesiastical world.

Some Egyptian temples took many tens of years to build; the obelisks, all in single blocks, were brought for hundreds of miles down the Nile. The building of a solar temple like that of Amen-Rā meant to the Egyptians a very serious undertaking indeed.

Some of the structural details are of a very curious nature, while the general arrangement of the temple itself is no less extraordinary. First, with regard to the temple axis. It seems to be a general rule that from the entrance-pylon the temple stretches through various halls of different sizes and details, until at last, at the extreme end, what is called the Sanctuary, Naos, Adytum, or Holy of Holies, is reached. The end of the temple at which the pylons are situated is open, the other is closed. These lofty pylons, and even the walls, are sometimes covered with the most wonderful drawings and hieroglyphic figures and records. Stretching in front of the pylons, extending sometimes very far in front, are rows of sphinxes. This principle is carried to such an extent that in some cases separate isolated gates have been built right in front and exactly in the alignment of the temple.

From one end of the temple to the other we find the axis marked out by narrow apertures in the various pylons, and many walls with doors crossing the axis.

VIEW TO THE SOUTH-WEST FROM THE SACRED LAKE OF AMEN-RĀ.

In the temple of Amen-Rā there are 17 or 18 of these apertures, limiting the light which falls into the Holy of Holies or the Sanctuary. This construction gives one a very

RUINS OF DOOR AT ENTRANCE OF THE SANCTUARY.

definite impression that every part of the temple was built to subserve a special object, viz., to limit the light which fell on its front into a narrow beam, and to carry it to the other extremity of the temple—into the sanctuary, so that once a

THE OBELISKS NEAR THE OLDEST PART OF THE TEMPLE OF AMEN-RĀ AT KARNAK.

year when the sun set at the solstice the light passed without interruption along the whole length of the temple, finally illuminating the Sanctuary in most resplendent fashion and striking the Sanctuary wall. The wall of the Sanctuary opposite to the entrance of the temple was always blocked. There is no case in which the beam of light can pass absolutely through the temple.

The point was to provide an axis open at one end and absolutely closed at the other, the open courts being only found towards that end towards which the temple opened, the other end being all but absolutely dark and quite blocked up at the extremity.

These sunlight effects were fully appreciated. Referring to the obelisks erected by Queen Hātshepset as a monument to her father Amen, an inscription at the base of one of these says, "They are seen an endless number of miles off: *it is a flood of shining splendour when the sun shines between the two*;"[1] and again, "The sun's disc shines between them as when it rises from the horizon of heaven."[2]

Passing from the temple at Karnak to others in a better state of preservation, we can gather that the part of the axis furthest from the entrance was covered, so that in the *penetralia* there was only a dim religious light. The entrance is also, as it were, guarded by a massive exterior pylon, as in the more or less modern temple of Edfû. This, again, reduces the light in the interior.

It is easy to recognise that these arrangements bear out the idea of an astronomical use of the temple.

First of all we know that the temple was directed to the place of the sun's setting; and if the Egyptians wished to lead

[1] Brugsch, "Egypt," p. 174.

[2] Inscription south side of obelisk quoted in "Records of the Past," Vol. XII. (Letter from Capt. Lyons).

INNER COURT AND SANCTUARY AT EDFÛ.

(*From a Photograph by the Author.*)

the narrow shaft of light which was bound to enter the temple, since it was directed to the sunset, they would have contrived the very system of gradually narrowing doors which we have found to be one of the special features of the temple.

The doors were considered as very important—and no wonder. In the account given of Thothmes III.'s restoration of the temple of Amen-Rā, we read that after the building had been constructed in a "position corresponding to the four quarters of heaven" the great stone gateways were erected.

"The first had doors of real acacia wood covered with plates of gold, fastened with black bronze and iron."

Then came a propylon (Bekhen) with three other gates connected with it covered with plates of copper, and the sacrifices were brought through these.[1]

This idea is strengthened by considering the construction of the astronomical telescope. Although the Egyptians knew nothing about telescopes, it would seem that they had the same problem before them which we solve by a special arrangement in the modern telescope—they wanted to keep the light pure, and to lead it into their sanctuary as we lead it to the eyepiece. To keep the light that passes into the eyepiece of a modern telescope pure, we have between the object-glass and the eyepiece a series of what are called diaphragms; that is, a series of rings right along the tube, the inner diameters of the rings being greatest close to the object-glass, and smallest close to the eyepiece; these diaphragms must so be made that all the light from the object-glass shall fall upon the eyepiece, without loss or reflection by the tube.

These apertures in the pylons and separating walls of Egyptian temples exactly represent the diaphragms in the modern telescope.

[1] Brugsch, "Egypt," p. 177.

What then was the real use of these pylons and these diaphragms? It was to keep all stray light out of the carefully roofed and darkened Sanctuary; but why was the Sanctuary to be kept in darkness?

The first point that I wish to make is that these temples—whatever view may be entertained with regard to their worship or the ceremonial in them—were undoubtedly constructed among other reasons for the purpose of obtaining an exact observation of the precise time of the solstice. The priests having this power at their disposal, would not be likely to neglect it, for they ruled by knowledge. The temples were, then, astronomical observatories, and the first observatories that we know of in the world.

If we consider them as horizontal telescopes used for the purpose I have suggested, we at once understand the long axis, and the series of gradually narrowing diaphragms, for the longer the beam of light used the greater is the accuracy that can be obtained.

Independently of ceremonial reasons—there is a good deal to be said under that head—it is quite clear that the darker the sanctuary the more obvious will be the patch of light on the end wall, and the more easily can its position be located. It was important to do this on the two or three days near the solstice, in order to get an idea of the exact time at which the solstice took place. We find that a narrow beam of sunlight coming through a narrow entrance some 500 yards away from the door of the Holy of Holies would, provided the temple were properly orientated to the solstice, and provided the solstice occurred at the absolute moment of sunrise or sunset according to which the temple was being utilised, practically flash into the sanctuary and remain there for about a couple of minutes, and then pass away. The flash would be a crescendo and diminuendo, but the whole thing would not last above two minutes or thereabouts, and might be considerably reduced by an arrangement of curtains. Supposing the solstice did not occur at the precise moment of sunrise or sunset, and provided the Egyptians by any means whatever were able to divide the days and the nights into more or less equal intervals of time, two or three observations of the sun rising at the solstice on three different mornings, or of the sunset at the solstice on three different evenings, would enable a careful observer to say whether the solstice had occurred at the exact moment of sunrise or sunset, or at some interval between two successive sunrises or sunsets, and what that interval was.

We may conclude that there was some purpose of utility to be served, and the solar temples could have been used undoubtedly, among other things, for determining the exact length of the solar year.

I now come to my next point, which is that here we have the true origin of our present means of measuring time; that our year as we know it was first determined in these Egyptian temples and by the Egyptians. The magnificent burst of the light at sunset into the sanctuary would show that a new true solar year was beginning. It so happens that the summer solstice was the time when the Nile began, and still begins, to rise; so that in Egypt the priests were enabled to determine, year after year, not only the length of the year, but the exact time of its commencement. This, however, they apparently kept to themselves, for the year in use, called the vague year, began at different times of the true year through a long cycle, as I shall show in subsequent chapters.

If the Egyptians wished to use the temple for ceremonial purposes, the magnificent beam of light thrown into the temple at the sunset hour would give them opportunities and even suggestions for so doing; for instance, they might place an image of the god in the sanctuary and allow the light to flash upon it. We should have a "manifestation of Rā" with a vengeance during the brief time the white flood of sunlight fell on it; be it remembered that in the dry and clear air of Egypt the sun casts a shadow five seconds after the first little point of it has been seen above the horizon. So that at sunrise and sunset in Egypt the light is very strong, and not tempered as with us. They did this: we not only find the exact allocation of words "the manifestation of Rā," but what happened is described. One of the inscriptions relating to the manifestation of Rā has been translated by De Rougé as follows:—

> "Il vint en passant vers le temple de Rā; il entra dans le temple en adorant (deux fois). Le χer-heb [celebrant] invoqua (celui qui) repousse les plaies du roi; il remplit les rites de la porte; il prit le seteb, il se purifia par l'encens; il fit une libation; il apporta les fleurs de *Habenben* [a part of the temple]; il apporta le parfum (?). Il monta les degrés vers l'adytum grand, pour voir Rā dans Habenben; lui-même se tint seul; il poussa le verrou; il ouvrit les portes; il vit son père Rā dans Habenben; il vénéra la barque de Rā et la barque de Tum. Il tira les portes, et posa la terre sigillaire (qu'il) scella avec le sceau du roi. Lui-même ordonne aux prêtres, 'J'ai placé le sceau; que n'entre pas quelqu'un dedans de tout roi qui se tiendra (là).'"[1]

In the quotation the apparatus of doors is referred to, and it is not difficult to understand that by a particular arrangement of them it would be easily possible to allow the flash which lighted up the image of the god to be of very brief duration. Remember that the sanctuary was dark, that the king stood with his back to the pylon (and therefore to the sun). Under these circumstances, to an excited imagination it would be the god himself and not his image which appeared. Maspero[2] adduces much evidence to show that the priests were not

[1] "Chrestomathie Égyptienne," De Rougé, iii., p. 60.
[2] "Egyptian Archæology," English edition, p. 105.

above pious frauds even in the worship connected with the Holy of Holies:—

"The shrines [in the sanctuary] are little chapels of wood or stone, in which the spirit of the deity was supposed at all times to dwell, and which on ceremonial occasions contained his image. The sacred barks were built after the model of the Bari, or boat in which the sun performed his daily course. The shrine was placed amidships of the boat and covered with a veil or curtain, to conceal its contents from all spectators. . . . . We have not as yet discovered any of the statues employed in the ceremonial, but we know what they were like, what part they played, and of what materials they were made. They were animated. . . . . They spoke, moved, acted—not metaphorically, but actually. . . . . Interminable avenues of sphinxes, gigantic obelisks, massive pylons, halls of a hundred columns, mysterious chambers of perpetual night—in a word, the whole Egyptian Temple and its dependencies were built by way of a hiding-place for a performing puppet, of which the wires were worked by a priest."

In an inscription which covers, according to Brugsch, an entire wall near the Holy of Holies in the temple of Amen-Rā it is stated that a beautiful harp, inlaid with silver and gold and precious stones, on which to sing the praises of the god, statues of the god himself, and numerous gates (Selkhet) with locks of copper and dark bronze, to protect the Holy of Holies from intrusion, were among the gifts to the priests.[1]

Thothmes III., in his account of his embellishments at Karnak, says of the statues of the gods and of their secret place (possibly the Adytum) that they were "more glorious than what is created in heaven, more secret than the place of the abyss, and more [invisible] than what is in the ocean."[2]

[1] Brugsch, "Egypt," p. 174.

[2] Brugsch, *op. cit.*, p. 187.

CHAPTER XI.

THE AGE OF THE TEMPLE OF AMEN-RĀ AT KARNAK.

If it be accepted that the arguments already put forward justify us in regarding the temple of Amen-Rā as a solstitial solar temple, we are brought face to face with the fact that if it be of any great antiquity its orientation should be such that it will no longer receive the light of the setting sun at the summer solstice along its axis.

This results from the fact that there is a slow change in what is called the obliquity of the ecliptic—that is, the angle between the plane of the earth's equator and the plane of the ecliptic; this change is brought about by the attraction of the other planetary bodies affecting the plane of the ecliptic. If these planes approach each other, the obliquity will be reduced; the present obliquity is something like 23° 27′; we know that 5,000 B.C. it was 24° 22′, nearly a degree more. A difference of 1° means, then, a difference of time of about seven thousand years. It may go down to something below 21°. Since the obliquity has been decreasing for many thousand years, a temple directed to the rising or setting sun at the solstice some thousands of years ago had a greater amplitude than it requires now.

It will be readily understood that if the orientation of the temple and the height of the hills towards which it points be accurately known, knowing also the precise obliquity of the ecliptic at different epochs, we have an astronomical means of determining the date of the original foundation of the temple, supposing, of course, that it was founded to observe the solstice.

But before I go into these matters it is essential that the evidence of Egyptologists should be considered. Very fortunately for us in these inquiries the temple of Amen-Rā is one of those most carefully studied by Mariette, so that the *dernier mot* of the archæologist is at our disposal.

Mariette, in his magnificent memoir on Karnak,[1] surpassed himself in the care and sagacity which he displayed in endeavouring to fix dates for the various structures in that wonderful temple-field, and among them the various parts of the temple of Amen-Rā.

In his maps, to which I now refer, each part of the temple is coloured according to the supposed date of its building. He points out first of all that the inscriptions on the walls must be disregarded, as they could have been put there at any date after the temples were built. On this point I quote Mariette's own words:—[2]

"Les couleurs marquées sur le plan servent à indiquer, au moyen de la légende explicative placée en marge, les époques diverses de la construction des temples et de leurs parties. Quelques mots d'explication sont ici nécessaires. Un mur porte les cartouches de Menephtah ; mais il peut avoir été construit deux-cent-cinquante ans plus tôt par Thoutmès III. Les époques de la décoration ne sont ainsi pas toujours les époques de la construction. Pour avoir les époques de la décoration, il ne s'agit que de regarder les murs et les inscriptions dont ils sont couverts. Pour avoir les époques de la construction, tout un travail de confrontation, de comparaison, est nécessaire. Il faut s'assurer si les mêmes mains qui ont construit le mur l'ont décoré ; dans le cas contraire, il faut faire intervenir l'archéologie dans toutes les branches de cette science qui touchent à l'observation des lieux, au mode de construction, à l'agencement des pierres, au choix et à l'appareillage des matériaux."

Taking the temple in its generality, he finds that, so far as his inquiries had carried him, parts were certainly built at a

[1] "Karnak. Étude topographique et archéologique."

[2] Mariette, *op. cit.*, text, p. 2.

time as ancient as the twelfth dynasty—say 2400 or 3000 B.C., according to the authority in these matters that we may prefer.

Then again we have dates given and indications of kings through the eighteenth and nineteenth dynasties, and then again on to the times of the Ptolemies.

In such an inquiry we must have archæological dates on which we can rely. In the date assigned to the time of Mena by various Egyptologists we find a difference of nearly—in fact, rather more than—a thousand years in our authorities. In the twelfth dynasty we find a difference of five hundred years; but in the later dynasties, such as the eighteenth, the difference is reduced in some cases to ten years or so. So that in the later dynasties we know pretty well what time is in question. We are therefore on firm ground.

The first point to which I wish to call attention is that according to Mariette the building dates change along the open axis of the temple. From photographs I took when in Egypt I found reason to believe that the direction of the axis has been slightly changed at the west end.

If we refer to the plan of the temple, the point of importance to us in our present inquiry has relation to the circumstances connected with the buildings of the temple itself. We have in the outer court to the north-west certain pillars which were built by one of the Ethiopian kings. These I mark **I,I** (see page 118). There is the temple **M,** built by Rameses III., according to Mariette. There are walls with columns, marked **2,2,** built by the twenty-second dynasty, north and south of this outer court; and then there is the temple **L** in the outer court, supposed to have been built by Seti II. The western part of the temple, therefore, is of no high antiquity. To find this we have to go some 200 yards to the south-east. Near the central portion of the temple (marked 4) there are traces of the twelfth or possibly the eleventh dynasty. What existed then might have been a shrine with nothing to the north-west or south-east of it.

This seems almost to have been its condition at the time of Thothmes III. even.

According to an inscription quoted by Brugsch,[1] "The king (Thothmes III.) found it in the form of a brick building, in a very dilapidated condition, being a work of his predecessors. The king with his own hand performed the solemn laying of the foundation stone for this monument."

From this point, indeed, the temple seems to have extended in both directions—that is, north-west and south-east—the sanctuary being thrown back to the eastward and pylons added to the westward.

It follows from the above very brief sketch that the original orientation of the original shrine is to be gathered from the walls towards the centre of the present ruins.

Let us agree to this. The Egyptologist already gives us eleventh-dynasty time, say 2500 B.C. for a part of the existing temple.

Let us now pass to the astronomical problem. Lepsius and others have measured the amplitude of this part of the temple. It is given as 26° or 26° 30′ N. of W.

When there I measured the height of the opposite hills (near the tombs of the kings) roughly at $2\frac{1}{2}°$. If we, therefore, deal with the amplitude, considering the height of the hills as $2\frac{1}{2}°$, we find that, as the horizon was above the sea horizon and the sun travels down an inclined path from south to north, it would meet the hill sooner than the sea horizon; the apparent amplitude would, therefore, be less than the true one, so that

[1] "History" p. 175.

we get an amplitude of 25° instead of 26°, and if we correct that for refraction we get 25½°.

Let us take the lower amplitudes. We can construct the following table:—

With present obliquity 23° 30′ we have at Thebes, lat. 25° 40′ amplitude on horizon (sun's centre)	26°
Corrected for refraction	26° 30′
The amplitude behind hill, 2½° high, will be	25°
Making correction for refraction	25° 5′

So that, taking the lowest amplitude, the temple axis points almost 1° too much to the north.

I have already mentioned that the photographs I had taken of the temple axis towards, and from the outside of, the Ptolemaic pylon indicated a twist in the temple axis. This was a question that in the absence of accurate measurements could only be determined by an actual observation of the solstice.

This being so, I begged the intervention of Col. Sir Colin Scott-Moncrieff, the Under Secretary of State of the Public Works Department in Egypt, to detail one of his officers to make observations of the summer solstice of 1891. He was good enough to accede to my request, and I proceed to give extracts from the report of the officer in question, Mr. P. J. G. Wakefield, to Mr. Allan Joseph, the Director of Works and Irrigation:—

"In accordance with instructions received, I made the following observations at Karnak on June 21st, 1891:—

"I found that the points which I have marked A, C, D on the photographic plan (being the centres of the Pylon of Rameses I., the Pylon of Thothmes I., and the shrine or sanctuary of Philip III. of Macedon (?) respectively) were all in a straight line. B is a point midway between the only two opposite pillars of which the bases are intact (one set up by Rameses I., and the other by Seti I.), and was very nearly in line; probably the true centre between the pillars (which is difficult to obtain) would be exactly so. The centre of the Great Pylon (Ptolemaic) is not in line at all with these points, there being 1° difference

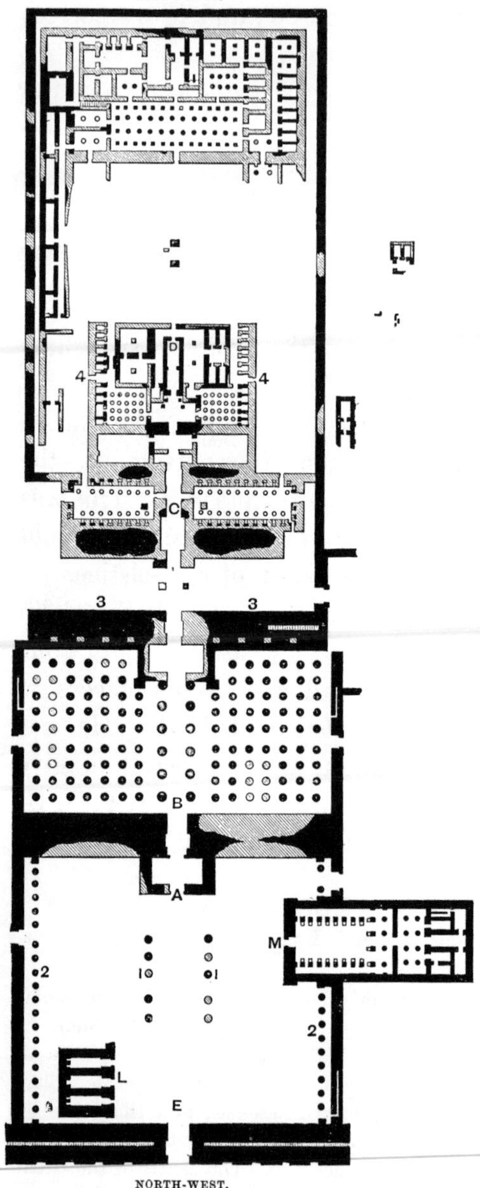

PLAN OF THE TEMPLE OF AMEN-RĀ, SHOWING THE POINTS REFERRED TO IN THE PRESENT CHAPTER.

between D A prolonged and A E; I therefore accepted the line D C A as the true axis.

"From an inspection made on June 20th, it appeared to me that the setting sun would not be visible from any of the points indicated by Professor Norman Lockyer. I therefore placed the theodolite at A. I regret to say that my above supposition was correct, as even from A I was only able to see a portion of the setting sun, the remainder being hidden behind the south wall of the Great Pylon. I obtained, however, one reading, the right limb at, as nearly as I could judge, the moment of impact of the sun's diameter with the hill."

Of the measures given the most important are the angle between the axis of the temple looking south-east from A and the north point 116° 23′ 40″ (amplitude 26° 23′) and the angle between the top of the hills and the horizontal 2° 36′ 20″.[1] These measures, therefore, entirely justified the result of the calculations I have before given,

[1] Nissen in his important memoir does not refer to this hill; his conclusions, therefore, are not absolutely justified by the facts as he states them.

and prove that the interval of over 5000 years is sufficient to cause us to detect the change in the obliquity of the ecliptic by this method of observing the sunset at a solstice with an instrument built on so large a scale.

Taking the orientation as 26°, and taking hills and refraction into consideration, we find that the true horizon sunset amplitude would be 27° 30′. This amplitude gives us for Thebes a declination of 24° 18′.

This was the obliquity of the ecliptic in the year 3700 B.C., and this is therefore the date of the *foundation* of the shrine of Amen-Rā at Karnak, so far as we can determine it astronomically with the available data; but about these there is still an element of doubt, for, so far as I learn, the recent magnetic readings have not been checked by astronomical observations.

CHAPTER XII.

THE STARS—THEIR RISINGS AND SETTINGS.

From what has been stated it is not too much to assume that the Egyptians observed, and taught people to observe, the sun on the horizon.

This being so, the chances are that at first they would observe the stars on the horizon too, both stars rising and stars setting; this indeed is rendered more probable by the very careful way in which early astronomers defined the various conditions under which a star can rise or set, always, be it well remembered, in relation to the sun.

It must not be forgotten that the ancients had no telescopes, and had to use their horizon as the only scientific instrument which they possessed. They spoke of a star as rising or setting cosmically, achronically, or heliacally.

The cosmic rising meant that the star rose, and the cosmic setting meant that the star set, at the same moment as the sun—that is, that along the eastern horizon we should see the star rising at the moment of sunrise, or along the western horizon a star setting at the moment of the sun setting; but unless certain very obvious precautions were taken it is clear that neither the rising nor the setting star would be seen, in consequence of the presence of daylight. The achronical rising or setting is different from the cosmic in this respect—that we have the star rising when the sun is setting, or setting when the sun is rising. Finally we have the heliacal rising and setting; that is taken to be that the star appeared in the morning a little in advance of the sunrise, or set at twilight a little later than the sun.

It is quite clear that if we observe a star rising in the dawn, it will get more and more difficult to observe the nearer the time of sunrise is approached. Therefore, what the ancients did was to determine a time before sunrise in the early dawn at which the star could be very obviously and clearly seen to rise. The term "heliacal rising" was coined to represent a star rising visibly in the dawn, therefore, before the sun. Generally throughout Egypt the sun was supposed to be something like 10° below the horizon when a star was stated to rise *heliacally*.

The following table from Biot should make matters quite clear :—

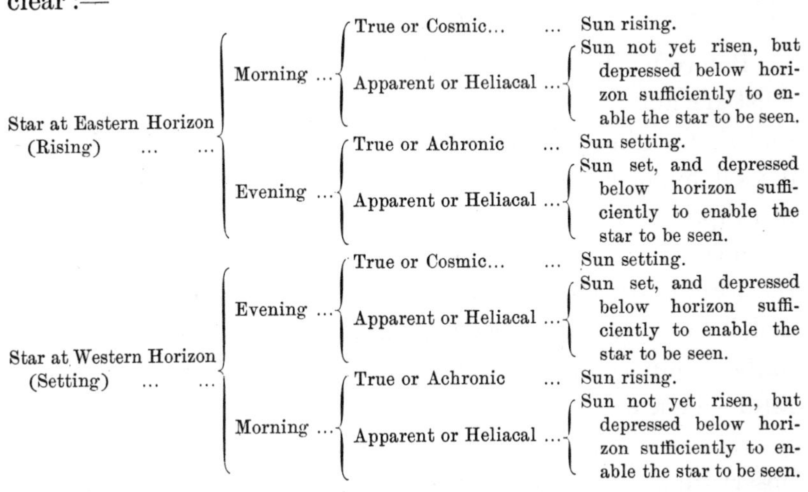

Star at Eastern Horizon (Rising)	Morning ...	True or Cosmic... ...	Sun rising.
		Apparent or Heliacal ...	Sun not yet risen, but depressed below horizon sufficiently to enable the star to be seen.
	Evening ...	True or Achronic ...	Sun setting.
		Apparent or Heliacal ...	Sun set, and depressed below horizon sufficiently to enable the star to be seen.
Star at Western Horizon (Setting)	Evening ...	True or Cosmic... ...	Sun setting.
		Apparent or Heliacal ...	Sun set, and depressed below horizon sufficiently to enable the star to be seen.
	Morning ...	True or Achronic ...	Sun rising.
		Apparent or Heliacal ...	Sun not yet risen, but depressed below horizon sufficiently to enable the star to be seen.

It is Ideler's opinion that, in Ptolemy's time, in the case of stars of the first magnitude, for heliacal risings and settings, if the star and sun were on the same horizon, a depression of the sun of 11° was taken; if on opposite horizons, a depression of 7°. For stars of the second magnitude these values were

14° and $8\frac{1}{2}$°. But if temples were employed as I have suggested, even cosmic and achronic risings and settings could be observed in the case of the brightest stars.

But it must not be imagined that, even in Egypt, all stars can be observed the moment they are above the horizon. In the morning, especially, there are mists, so that all but the brightest stars are often invisible till they are 1° or 2° high. On this point I quote Biot:—

"Comme le rapporte Nouet, l'astronome de l'expédition française, on n'y aperçoit jamais à leur lever les étoiles de 2° et de 3° grandeur même dans les plus belles nuits, à cause d'une bande constant de vapeurs qui borde l'horizon.[1] Aussi en expliquant le calcul des levers héliaques dans l'Almageste, Ptolémée a-t-il soin de remarquer[2] que les annonces qu'on voudrait faire de ces phénomènes seront toujours très-incertaines, à cause de l'état des couches d'air dans lesquelles on les observe, et à cause de la difficulté optique qu'on éprouve à saisir la première apparition, comme il dit lui-même en avoir fait l'expérience."[3]

Before we begin to consider the question of stars at all, we must be able to describe them—to speak of them in a way that shall define exactly which star is meant. We can in these days define a star according to its constellation, or its equatorial or ecliptic co-ordinates, but all these means of reference were unknown to the earliest observers. Still we may assume that the Egyptians could define some of the stars in some fashion; and it is evident that we here approach a matter of the very highest importance for our subject, to which I shall have to return in a subsequent chapter.

So far as we have been dealing with the sun and the observations of the sun at rising and setting, we have taken for granted that the amplitude of the sun at the solstices does not change; the amplitude of 26° at Thebes for the solstices is

[1] "Œuvres de Volney," vol. v., p. 431.

[2] "Ptolemy Almagest VIII.," chap. vi.

[3] "Recherches sur l'année vague des Egyptiens," by M. Biot, Académie des Sciences 4th April, 1831.

practically, though as we have seen not absolutely, invariable for a thousand years; but one of the results of astronomical work is that the *stars* are known to behave quite differently. In consequence of what is called *precession* the stars change their place with regard to the pole of the equator; and further, in consequence of this movement, the position of the sun among the stars at the solstices and equinoxes changes also.

In reference to the sun's path we considered what are called the ecliptic and the equatorial co-ordinates. The ecliptic defines the plane in which the earth moves round the sun, and 90° from that plane we have the pole of the heavens; celestial latitude we found reckoned from the plane of the ecliptic north and south up to the pole of the heavens, and celestial longitude was reckoned along the plane of the ecliptic from the first point of Aries. We had also declination reckoned from the equator of the earth prolonged to the stars, and right ascension reckoned along the equator from the first point of Aries.

The pole of the heavens or of the ecliptic, then, we must regard as practically, but not absolutely, fixed; but the pole of the earth's equator is not fixed, it slowly moves round the pole of the heavens. *In consequence of that movement there is a change of declination in a star's place.*

Going back to the diagram (p. 49), we find that the amplitude of a body rising or setting at Thebes or anywhere else depends upon its declination; so that if from any cause the declination of a star changes, its amplitude must change.

That is the first point where we meet with difficulty, because if the amplitude changes it is the same as saying that the place of star-rising or star-setting changes; that is, a star which rises in the east in a certain amplitude this year will change its amplitude at some future time.

In the last chapter I referred to one of the difficulties

of modern inquiries into the orientation of ancient temples, which arises from the fact that the sun has not always, at the solstices, risen or set at exactly the same points of the horizon. We now find ourselves face to face with the fact that the stars do not rise or set at the same points century after century. We saw that the change in the position of the sun on the horizon at the solstices is due to a very small change of obliquity of the ecliptic, so that in a matter of something like 6,000 years the position of the sun at sunrise and sunset on the horizon may be varied by, roughly speaking, 1 degree. But in the case of the stars the matter is very much more serious, because in the course of something like 13,000 years the rising- or setting-places of a star may vary by something like 47° along the horizon north or south.

So that in the cases both of sun and stars there is no real fixity in the places of rising or setting, although of course those who made the first observations and built the first temples were not in a position to know this.

The real cause of this precessional movement which causes the stars to change their places lies in the fact that the earth is not a sphere, its equatorial diameter being longer than its polar diameter, so that there is a mass of matter round the equator in excess of what we should get if the earth were spherical. Suppose that matter to be represented by a ring. The ring is differently presented to the sun, one part being nearer than the other, the nearer part being attracted more forcibly. If we take the point in the ring nearest the sun where there is the greatest attraction, and draw a line to the opposite point where the attraction is least, we can show that the case stands in this way: the sun's pull may be analysed into two forces, one of them represented by the line joining the centre of the sun and the centre of the ring, and another at right angles to it let fall from the point most strongly attracted on to the first line. The question is, what will that force at right angles do?

The figure below represents a model illustrating the rotation of the earth on its axis, and the concurrent revolution of the sun round the earth once a year. To represent the downward

MODEL ILLUSTRATING THE PRECESSION OF THE EQUINOXES.

force it is perfectly fair if I add a weight. The moment this is done the axis of the gyroscope representing the earth's axis, instead of retaining its direction to the same point as it did before, now describes a circle round the pole of the heavens.

It is now a recognised principle that there is, so to speak, a wobble of the earth's axis round the pole of the heavens, in consequence of the attraction of the sun on the nearer point of this equatorial ring being greater than on the part of the ring further removed from it. That precessional movement is not quite so simple as it is shown by the model, because what the sun does in this way is done to a very much larger extent by the moon, the moon being so very much nearer to us.

In consequence, then, of this luni-solar precession we have a variation of the points of intersection of the planes of the earth's equator and of the ecliptic; in consequence of that we have a difference in the constellations in which the sun is at the time of the solstices and the equinoxes; and, still more important from our present point of view, we have another difference, viz., that the declinations, and therefore the amplitudes, and therefore the places of setting and rising of the stars, change from century to century.

Now that we have thus become acquainted with the physical cause of that movement of the earth's axis which gives rise to what is called the precession of the equinoxes, we have next to enter with somewhat greater detail into some of the results of the movement.

The change of direction of the axis in space has a cycle of something between 25,000 and 26,000 years. As it is a question of the change of the position of the celestial equator, or rather of the pole of the celestial equator, amongst the stars in relation to the pole of the heavens, of course the declinations of stars will be changed to a very considerable extent; indeed, we have seen that the declination of a star can vary by twice the amount of the obliquity, or say 47°, so that a star at one time may have zero declination—that is, it may lie on the equator—and at another it may have a declination 47° N. or S. Or, again, a star may be the pole star at one particular time, and at another it will be distant from the pole no less than 47°. Although we get this enormous change in one equatorial co-ordinate, there would from this cause alone be practically no change with regard to the corresponding ecliptic co-ordinate—that is to say, the position of the star with reference to the earth's movement round the sun. This movement takes place quite independently of the direction of the axis, so that while we get this tremendous swirl in declination, the latitudes of the stars or their distances from the ecliptic north or south will scarcely change at all.

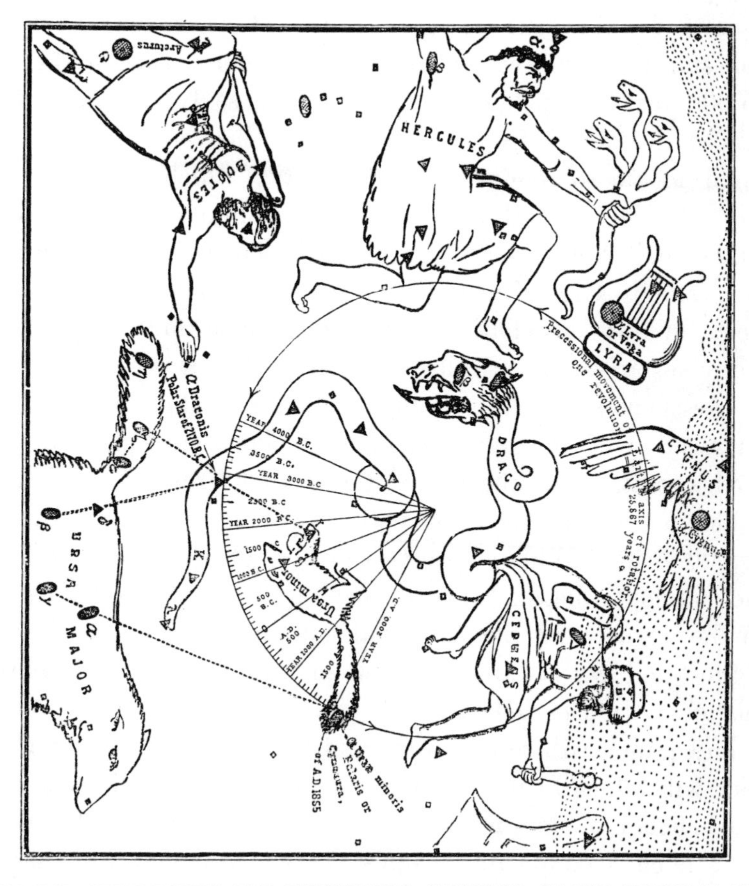

STAR-MAP, REPRESENTING THE PRECESSIONAL MOVEMENT OF THE CELESTIAL POLE FROM THE YEAR 4000 B.C. TO THE YEAR 2000 A.D. (*From Piazzi Smyth.*)

Symbols adopted to represent the magnitudes or brightnesses of the stars, 1st ●; 2nd ●; 3rd ▲; 4th ▪.

Among other important results of these movements dependent upon precession we have the various changes in the pole-star from period to period, due to the various positions occupied by the pole of the earth's equator. We thus see how in this period of 25,000 years or thereabouts the pole-stars will change, for a pole-star is merely the star near the pole of the equator for the time being. At present, as we all know, the pole-star is in the constellation Ursa Minor. During the last 25,000 years the pole-stars have been those lying nearest to a curved line struck from the pole of the heavens with a radius equal to the obliquity of the ecliptic, which, as we have seen, is liable to change within small limits; so that about 10,000 or 12,000 years ago the pole-star was no longer the little star in Ursa Minor that we all know, but the bright star Vega, in the constellation Lyra. Of course 25,000 years ago the pole-star was practically the same as it is at present.

Associated with this change in the pole-star, the point of intersection of the two fundamental planes (the plane of the earth's rotation and the plane of the earth's revolution) will be liable to change, and the period will be the same—about 25,000 years. Where these two planes cut each other we have the equinoxes, because the intersection of the planes defines for us the vernal and the autumnal equinoxes; when the sun is highest and lowest half-way between these points we have the solstices. In a period of 25,000 years the star which is nearest to an equinox will return to it, and that which is nearest a solstice will return to it. During the period there will be a constant change of stars marking the equinoxes and the solstices.

The chief points in the sun's yearly path then will change among the stars in consequence of this precession. It is perfectly clear that if we have a means of calculating back the old positions of stars, and if we have any very old observations, we can help matters very much, because the old observations—if they were accurately made—would tell us that such and such a star rose with the sun at the solstice or at the equinox at some special point of ancient time. If it be possible to calculate the time at which the star occupied that position with regard to the sun, we have an astronomical means of determining the time, within a few years, at which that particular observation was made.

Fortunately, we have such a means of calculation, and it has been employed very extensively at different periods, chiefly by M. Biot in France, and quite recently by German astronomers, in calculating the positions of the stars from the present time to a period of 2000 years B.C. We can thus determine with a very high degree of accuracy the latitude, longitude, right ascension, declination, and the relation of the stars to an equinox, a solstice, or a pole, as far back as we choose. Since we have the planes of the equator and ecliptic cutting each other at different points in consequence of the cause which I have pointed out—the attraction of the sun and moon—we have a fixed equator and a variable equator depending upon that. In consequence of the attraction of the planets upon the earth, the plane of the ecliptic itself is not fixed, so that we have not only a variable equator, but also a variable ecliptic. What has been done in these calculations is to determine the relations and the results of these variations.

The calculations undertaken for the special purposes of this book will be referred to later.

A simpler, though not so accurate a method consists in the use of a precessional globe. In this we have two fixed points at the part of the globe representing the poles of the heavens, on which the globe may be rotated; when this is done the stars move absolutely without any reference to the earth or to the plane of the equator, but purely with reference to the ecliptic. We have, then, this globe quite independent of the earth's axis. How can we make it dependent upon the earth's axis? We have two brass circles at a distance of $23\frac{1}{2}°$ from each pole of the heavens (north and south); these represent the circle described by the pole of the earth in the period of 25,000 years. In these circles are forty-eight holes in which I can fix two additional clamping screws, and rotate the globe with respect to them by throwing out of gear the two points which produced the ecliptic revolution.

If I use that part of the brass circle which is occupied by our present pole-star, we get the apparent revolution of the heavens with the earth's axis pointing to the pole-star of to-day. If we wish to investigate the position of things, say 8,000 years ago, we bring the globe back again to its bearings, and then adjust the screws into the holes in the brass circles which are proper for that period. When we have the globe arranged to 6000 years B.C. (*i.e.*, 8,000 years ago), in order to determine the equator at that time all we have to do is to paint a line on the globe in some water-colour, by holding a camel's-hair pencil at the east or west point of the wooden horizon. That line represents the equator 8,000 years ago. Having that line, of course, the intersection of the equator with the ecliptic will give us the equinoxes, so that we may affix a wafer to represent the vernal equinox. Or if we take that part of the ecliptic which is nearest to the North Pole, and, therefore, the N. declination of which is greatest, viz., $23\frac{1}{2}°$ N., we have there the position of the sun at the summer solstice, and $23\frac{1}{2}°$ S. will give us the position of the sun at the winter solstice. So by means of such a globe as this it is possible to determine roughly the position of the equator among the stars, and note those four important points in the solar year, the two equinoxes and the two solstices. I have taken a period of 8,000 years, but I might just as easily have taken a greater or a smaller one. By means of this arrangement, therefore, we can determine within a very small degree of error, without any laborious calculations, the distance of a star north or south of the equator, *i.e.*, its declination, at any point of past or future time.

The positions thus found, say, for intervals of 500 years, may be plotted on a curve, so that we can, with a considerable amount of accuracy, obtain the star's place for any year. Thus the globe may be made to tell us that in the year 1000 A.D. the declination of Fomalhaut was 35° S., in 1000 B.C. it was 42°, in 2000 it was about 44°, in 4000 it was a little over 42° again, but in 6000 B.C. it had got up to about 33°, and in 8000 B.C. to about 22°.

The curve of Capella falls from 41° N. at 0 A.D. to 10° at 5500 B.C., sc we have in these 5500 years in the case of this star run through a large part of that variation to which I have drawn attention.

I have ascertained that the globe is a very good guide indeed within something like 1° of declination. Considering the difficulty of the determination of amplitudes in the case of buildings, it is clear that the globe may be utilised with advantage, at all events to obtain a first approximation.

CHAPTER XIII.

THE EGYPTIAN HEAVENS—THE ZODIACS OF DENDERAH.

WE can readily understand that in the very beginning of observations in all countries, the moment man began to observe anything, he took note of the stars, and as soon as he began to talk about them he must have started by defining, in some way or other, the particular stars he meant.

Observers would first consider the brightest stars, and separate them from the dimmer ones; they would then discuss the stars which never set, and separate them from those which did rise and set; then they would take the most striking configurations, whether large or small. They would naturally, in a Northern clime, choose out the constellation the Great Bear, or Orion, and for small groups the Pleiades. These would attract attention, and be named before anything else. Then, later on, it would be imperative, in order to connect their solar with their stellar observations, that they should name the stars which lay along the sun's path in the heavens, or those the rising of which heralded the sunrise at their festivals. They would confine their attention to a belt round the equator rather than consider the configuration of stars half-way between the equator and the north pole. In all countries—India, China, Babylonia, Egypt—they had eventually such a girdle round the heavens, called by different names in different countries, and the use of this girdle of stars, which sometimes consisted of twenty-eight stations, sometimes of twenty-seven, and sometimes of less, was to enable them to define the place of the sun, moon, or of any of the planets in relation to any of these stars.

Not very many years ago, when the literature of China and India was as a sealed book, and the hieroglyphics of Egypt and the wedges of Babylonia were still unread, we had to depend for the earliest traces of astronomical observation upon the literatures of Greece and Syria, and according to these sources the asterisms first specialised and named were as follows:—

The Great Bear	Job (xxxviii. 31), Homer.
Orion	Job (ix. 9), Homer, Hesiod.
Pleiades, Hyades	Job (xxxviii. 31), Homer, Hesiod.
Sirius and the Great Dog ...	Hesiod (viii.), the name; Homer called it the Star of Autumn.
Aldebaran, the Bull	Homer, Hesiod.
Arcturus	Job (ix. 9; xxxviii. 32), Homer, Hesiod.
The Little Bear	Thales, Eudoxus, Aratus.
The Dragon	Eudoxus, Aratus.

In the Book of Job we read, "Canst thou bind the sweet influences of Pleiades, or loose the bands of Orion? Canst thou bring forth Mazzaroth in his season? or canst thou guide Arcturus with his sons?"

Here we have the difficulty which has met everybody in going back into these old records, because there was no absolute necessity for a common language at the time; it was open to everyone to call the stars any name they chose in any country, therefore it is difficult for scholars to find out what particular stars or constellations were meant by any particular words. In the Revised Version, Arcturus has given place to "the Bear with its train," and even our most distinguished scholars do not know what Mazzaroth means. I wrote to Professor Robertson Smith to ask him to give me the benefit of his great knowledge, and he tells me that Mazzaroth is probably that band of stars round the ecliptic or round the equator to which I have referred, but he will only commit himself to the statement that it is a probable enough conjecture; other people believe that it was a reference to the Milky Way.

I mention this to show how very difficult this inquiry really is. The "seven stars" are held by many to mean the Pleiades, and not the Great Bear; but this, I think, is very improbable.

Much is to be hoped from the study of the Babylonian records in relation to the Egyptian ones. This is a point I shall return to in the sequel.

In observing stars nowadays, we use a transit circle which is carried round by the earth so as to pick up the stars in different circles round the axis of the earth prolonged, and by altering the inclination of the telescope of this instrument we can first get a circle of one declination and then a circle of another.

The Egyptians did not usually employ meridian observations. Did the Egyptians make star maps? They certainly did, as we shall see.

The first bit of solid information specially bearing upon ancient Egyptian constellations was gained at the temple of Denderah, a place which the traveller up the Nile reaches before he arrives so far as Thebes. Perhaps among the reasons why so great attention was given to the so-called zodiacs of Denderah was the fact that one of them, having been rudely wrenched from its resting-place in the platform of one of the temples, had been carried to the museum in Paris, so that the thing itself was *en évidence* and capable of being examined by experts whose opinions were of value, and by all the world besides.

The chief temple, when explored by the French expedition,

was deeply buried in the sand. In the front part of it, covering the ceiling, before one enters the temple itself, there is displayed the square zodiac, so called, to which I shall have to refer briefly. The temple was pointed within a few degrees of north; at the north-east corner of the zodiac is a device, since found to represent the sunlight falling upon a statue of the Goddess of the Shrine. Investigations have shown that the zodiac includes a reference to a great many celestial phenomena of the utmost importance. There is no difficulty in recognising some of the zodiacal signs, but there the resemblance to the modern zodiac ends, for the reason that each of the strange processions of mythological personages represents not only constellations, with some of which we may be familiar, but a great deal more. It is noteworthy that the illustration of the very first astronomical point which we have to consider brings out the fact that it is impossible to disconnect Egyptian mythology from astronomy.

In the southern half of the zodiac, the lower part is occupied by stars represented in the guise of different mythological personages, sailing along in boats; and above them we get half of the zodiac with the signs of the Fish, the Ram, the Bull and the Twins represented. In the middle section the sun's course in different parts of the day, and different parts of the year, is given: whilst, outermost of all, we get the twelve solar positions, occupied by the sun each hour from rising to setting, represented by twelve boats. It may be here mentioned that in ancient Egypt, as in the modern Eastern world, both day and night always consisted of twelve hours; unequal, of course, the length of the hours varying according to the time of the year.

Now, if we take the opposite side, that is the north-west corner, we find that we have to do chiefly with the opposite

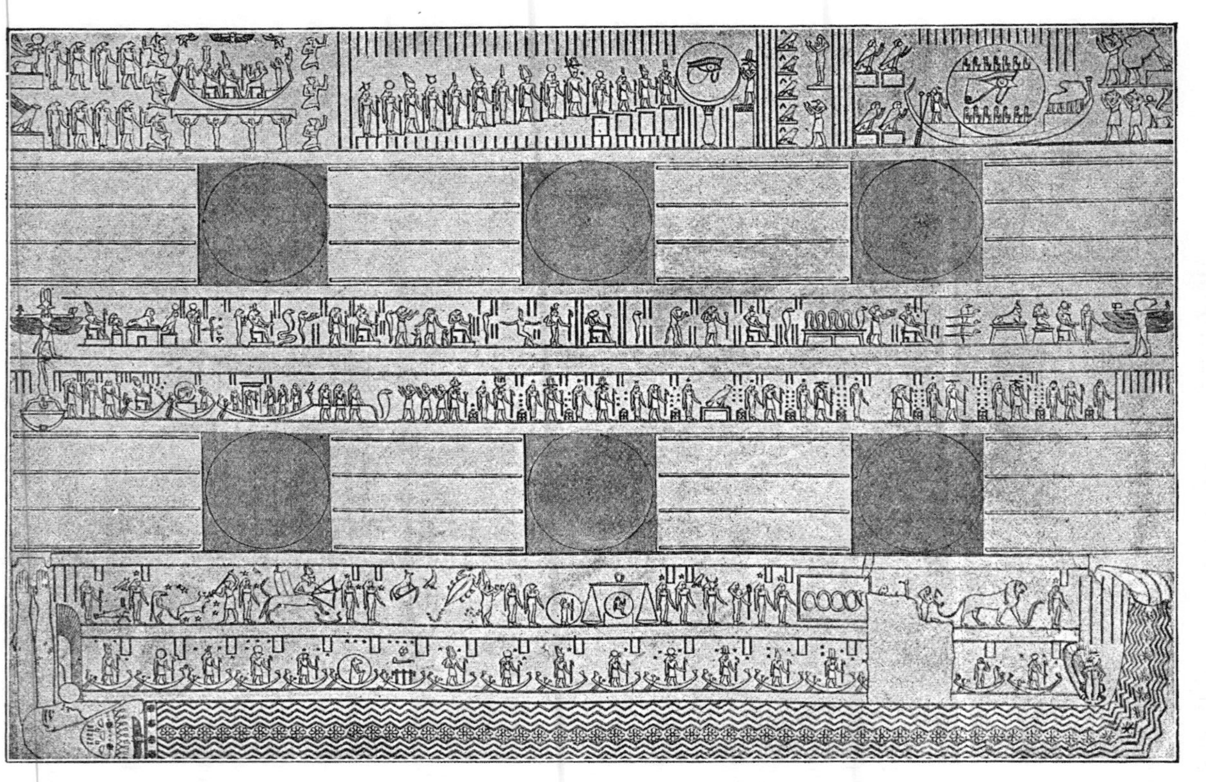

NORTHERN HALF OF THE SO-CALLED SQUARE ZODIAC OF DENDERAH.

part of the sky, including the signs of the Lion, the Scales, and Sagittarius, and below them other stars are represented as mythological personages in boats. The courses of the sun and moon are next given, and some of the lunar mythology is revealed to us. We see Osiris represented by the moon, and by an eye at the top of fourteen steps, which symbolise the fourteen days of the waxing moon.

In the square zodiac, then, there is an immense amount of astronomy. In the round zodiac, found in another temple (*see* p. 18), there are some points which at once claim our attention. There is, first, a mythological figure of a cow in a boat, and, near it, another mythological figure, which the subsequent reading of inscriptions has proved to represent the constellation Orion. In the centre of the zodiac we have a jackal, and there is very little doubt that it represents the constellation which we now call the Little Bear, which then, as now, was near the pole. Not far away, we get the leg of an animal; this, we now know, was a constellation called the Thigh, and there seems to be absolutely no question that it represents the constellation which we now call the Great Bear. Again, close by is another mythological form, which we know represents the Hippopotamus. This was made up out of some of the group of stars which forms the present constellation Draco. There are also two hieroglyphs which subsequent research has proved to represent setting stars and rising stars, so that, whatever may have been the date of this round zodiac of Denderah, it is clear that we are dealing with a time when the stars had been classed in constellations, one of which, the constellation Orion, even survives to our own day.

It is little to be wondered at that, when these revelations first burst upon the scientific world, great excitement was produced. It was obvious that we had to do with a nation which had very definite ideas of astronomy, and that the astronomy was very closely connected with worship. It was also certainly suggested by so many animal forms, that we had to do with a people whose condition was not unlike that of the American Indians—to take a well-known instance—at the beginning of this century, one in which each tribe, or clan, had chosen a special animal *totem*.

It so happened that, while these things were revealing themselves, the discussions concerning them, which took place among the scientific world of France, were partly influenced by the writings of a man of very brilliant imagination and of great erudition. I refer to Dupuis, according to whose views an almost fabulous antiquity might be assigned to ancient traditions in general and astronomical traditions in particular. It is needless to say, however, that there were others to take the extreme opposite view—who held the opinion that his imagination had run away with his learning.

With all this new work before them, and with a genius like Champollion's among them, it was not long before the French *savans* compelled the hieroglyphs to give up some of their secrets. First one word gave two or three letters, then another two or three more, and finally an alphabet and syllabary were constructed. So it was not long before some of the inscriptions at Denderah were read. Then it was found that the temple, as it then stood, had certainly been, partly at all events, embellished so late as the time of the Roman Emperors. Naturally there was then a tremendous reaction from the idea of fabulous antiquity which had been urged by the school of Dupuis. There were two radically opposed camps, led by Letronne, a distinguished archæologist, and Biot, one of the most eminent astronomers of his day, and both of these *savans* brought papers before the Academy of Inscriptions. Biot's first paper was read in 1822, and was replied to by Letronne in 1824; Biot wrote his next paper in 1844, in which he held to everything that he had stated in his first memoir; and this was replied to, the next year, by Letronne.

Biot had no difficulty whatever in arriving at the conclusion that, precisely as in the case of the sphere of Eudoxus, a prior bone of contention, however true it might be that

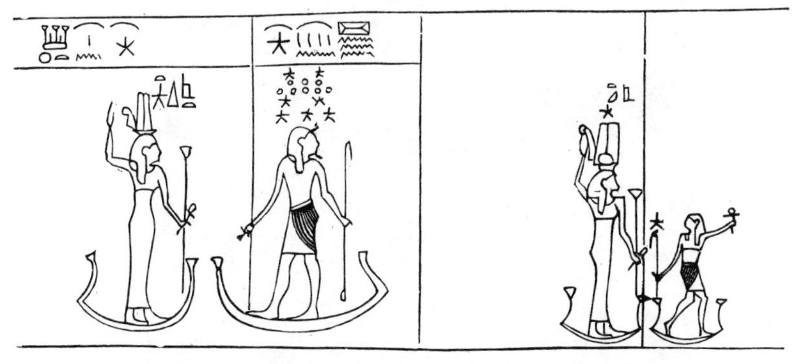

SIRIUS AND ORION (18TH DYNASTY). (*From Brugsch.*)

the circular zodiac had been sculptured in the time of the Roman emperors, still it certainly referred to a time far anterior; and he suggested that we have in it sculptures reproducing very old drawings, which had been made long before on parchment or on stone. He pointed out that in the condition of astronomy one would expect to be extant in ancient times, it was far easier to reproduce old drawings than to calculate back what the positions of the stars had been at some prior date, so that in his magnificent summing-up of the case in his last paper, he rested his scientific reputation on the statement that the sculptures of Denderah represent the celestial sphere on a plane round the north

ASTRONOMICAL DRAWINGS FROM BIBÂN EL-MULÛK (18TH DYNASTY). (*From "Description de l'Égypte."*)

pole of the equator at a year not far removed from 700 B.C. More than this, he stated that the time of the year was the time of the summer solstice, and the hour was midnight. He also showed that, calculating back what the position of the stars would have been at midnight on the 20th of June (Gregorian), 700 B.C., the constellations, and even many of the separate stars shown in the medallion, would occupy exactly the places they did occupy in the projection employed.

Let us then, for the moment, assume this to be true. What does it tell us? That 700 years B.C. in Egypt the solstice was recognised; a means of determining the instant of midnight with more or less precision was known; observations of the stars were regularly made; the risings of some of them were associated with the rising of the sun, and many of them had been collected into groups or constellations.

This is a wonderful result. I suppose that Biot is universally held to have proved his case; in fact, Brugsch, who is now regarded as one of the highest authorities in Egyptian history, has shown that almost every detail seen in the zodiac of Denderah reproduces inscriptions or astronomical figures, unearthed since the date of Biot's memoir, which, without doubt, must be referred to the time of the Eighteenth Dynasty—that is, 1700 B.C. or thereabouts; so that practically the Egyptologist has now chapter and verse for many things in the zodiac of Denderah dating 1,000 years before the period assigned to it by Biot.

The next point to notice is connected with the astronomical drawings which have been found in the Ramesseum at Thebes —drawings which also have very obvious connections with the zodiac of Denderah. On these we find the hieroglyphics for the different months—the constellations Orion, Hippopotamus, and Jackal, as we saw them at Denderah, and another form

RUINS OF THE RAMESSEUM, WHERE THE MONTH-TABLES WERE FOUND.

of the constellation of the Thigh. There is certainly the closest connection between the two sets of delineations.

Biot set himself to investigate what was the probable date to which the inscriptions in the Ramesseum referred. When we have the months arranged in a certain relationship to certain constellations we have an opening for the discussion of the precessional movements; in other words, for the consideration of the various changes brought about by the swinging of the pole of the equator round the pole of the ecliptic. Here, again, there was no uncertain sound given out by the research. Biot pointed out that we are here in presence of records, no longer of a summer solstice, as in the case of Denderah, but of a spring equinox, the date being 3285 B.C. He further suggested that, in all probability, one of the mythological figures might be a representation of the intersection of the ecliptic and the equator in the constellation Taurus at the date mentioned. This undoubtedly, to a large extent, justifies what Dupuis had long before pointed out—that the perpetual reference to the Bull found in ancient records and mythologies arose from the fact that this constellation occupied an important position at a critical time in the year, which would indicate a very considerable lapse of time. This idea was justified by the researches of Biot, because we are driven back by them to a date preceding 3000 years B.C. We find in the table at the Ramesseum distinct references to the Bull, the Lion and the Scorpion, and it is also clearly indicated that at that time the star Sirius rose heliacally at the beginning of the Nile-rise.

The month-table at Thebes tells us that the sun's journey in relation to some of the zodiacal constellations was perfectly familiar 5000 years ago.

CHAPTER XIV.

THE CIRCUMPOLAR CONSTELLATIONS: THE MYTH OF HORUS.

THERE was to all early peoples all the difference in the world, of course, between day and night, while we, with our firm knowledge, closely associate them. There was no artificial illumination such as we have, and the dark night did not so much typify rest as death; so that the coming of the glorious morning of tropical or sub-tropical climates seemed to be a re-awakening to all the joys and delights and activities of life; thus the difference between night and day was to the ancient Egyptians almost the difference between death and life. We can imagine that darkness thus considered by a mythologically-thinking people was regarded as the work of an enemy, and hence, in time, their natural enemies were represented as being the friends of darkness.

THE GOD OF DARKNESS—SET.

Here a very interesting astronomical point comes in. With these views, there must have been a very considerable difference in the way the Egyptians regarded those stars which were always visible and those which rose and set.

The region occupied by the stars always visible depends,

of course, upon the latitude of the place. Taking Thebes, with its latitude of 26°, as representing Egypt, the area of stars always visible was about one-fourth of that visible to us, so that there would be a very sharp distinction between the stars

VARIOUS FORMS OF ANUBIS.

constantly seen at night, and those which rose and set, the rising stars being regarded as heralds of the sunrise. It seems very probable that the circumpolar stars were quite early regarded as representing the powers of darkness, because they were there, visible in the dark, always disappearing and never appearing at sunrise. If that were so, no doubt prayers would be as necessary to propitiate them as those powers or gods which were more beneficent; and, as a matter of fact, one finds that the god Set—identified sometimes with Typhon, Anubis, and Tebha—was amongst the greatest gods of ancient Egypt.

The female form of Typhon—his wife—was called

Taurt or Thoueris, represented generally as a hippopotamus.

It is probable that the crocodile was a variant of the

FORMS OF TYPHON.

hippopotamus in some nomes, both having reference to our modern constellation Draco.

If we return for a moment to the zodiac of Denderah, we find that the constellations which I indicated—the Thigh, the Hippopotamus and the Jackal—represent our present constellations of the Great Bear, Draco, and the Little Bear, which were all of them circumpolar; that is, they neither rose nor set at the time of the inscription of the zodiac of Denderah. It therefore will not surprise us, with the above suggested explanation in mind, to hear that the Hippopotamus was called the Wife of Set, the Thigh the Thigh of Set, and the Jackal the Jackal of Set.

In the Book of the Dead, Chapter XVII., we read the

following reference to some of the northern stars and constellations:

"The gods Mestha, Hāpi, Tuamāutef, and Qebhsennuf are those, namely, which find themselves behind the constellation of the Thigh in the northern heavens."

Again, inscribed in the kings' graves at Thebes we read:

"The four Northern Genii are the four gods of the follower [some constellation]. They keep back the conflict of the terrible one [Typhon]. He is a great quarreller. They trim the foresail and look after the mizen in the bark of Rā, in company with the sailors, who are the four constellations[1] [aχemu-sek], which are found in the northern heavens. The constellation of the Thigh appears at the late rising. When this constellation is in the middle of the heavens, having

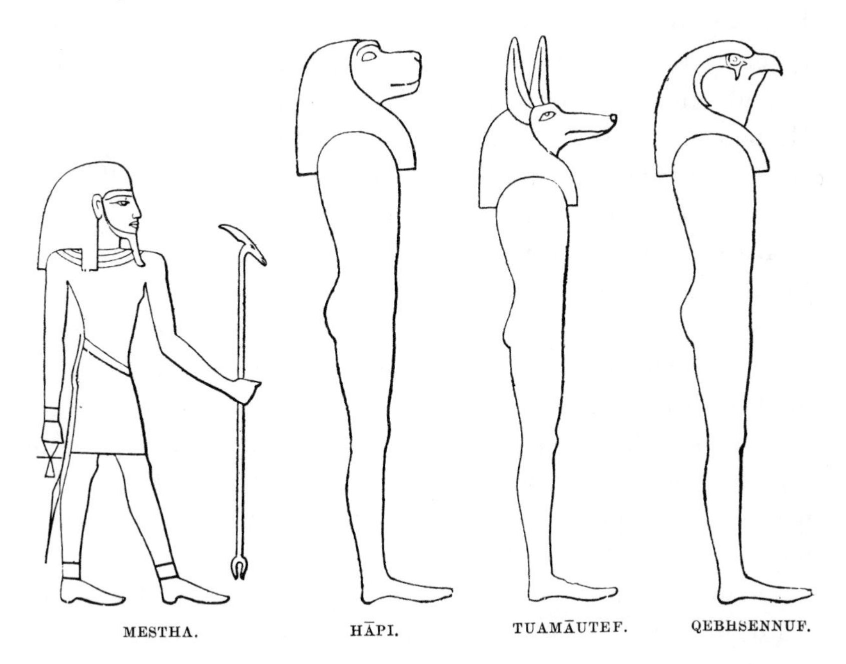

come to the south, where Orion lies [Orion typifying the southern part of the skies], the other stars are wending their way to the western horizon. Regarding the Thigh; it is the Thigh of Set, so long as it is seen in the northern heavens

[1] *I.e.*, the stars which never set.

there is a band [of stars ?] to the two [sword handles ?] in the shape of a great bronze chain. It is the place of Isis in the shape of a Hippopotamus to guard."

In the square zodiac at Denderah we find an illustration of the Hippopotamus and the Thigh, and the chain referred to in the inscription is there also. It will be quite worth while to see whether this chain is not justified by some line of stars between the chief stars in Draco and those in the Great Bear.

Let us now turn to the associated mythology. We see that the astronomical ideas have a most definite character; we learn also from the inscriptions dating from the Eighteenth Dynasty, that the Egyptians at that time recognised three different risings. There was the rising at sunset, the rising at midnight, and the rising at dawn. Plutarch says that the Hippopotamus was certainly one of the forms of Typhon, and a reference to the myth of Horus, so beautifully told twenty years ago and illustrated by Naville by the help of inscriptions at Edfû, will show how important this identification is.

Naville rightly pointed out how vital the study of mythology becomes with regard to the advancement of any kind of knowledge of the thoughts and actions of the ancient Egyptians. Mythology, as Bunsen said, is one of the poles of the existence of every nation; hence it will be well not to neglect the opportunity thus afforded of studying the astronomical basis of one of the best-known myths.

First a word about the mythology of Horus. Generally we begin with the statement usually made that Horus meant the young (or rising) sun. But inquiry shows that Horus was something more than this; the Egyptians were great generalisers.

If we put the facts already known into diagrammatic

form, we find that the condition of things is something like the following:—

HORUS = SUN, PLANET, or CONSTELLATION RISING.

SUN.	PLANETS.	CONSTELLATIONS.	
Horus	Mars as Hor-χuti (Laughing Horus) (Red Horus)	Orion Sah-Horus	Northern constellations Set-Horus.

The table shows that, although the Egyptians undoubtedly called the rising sun Horus, the planets and constellations when rising were in certain cases called Horus too. We do not get any individual star rising referred to as Horus; they were always considered as goddesses. Hence, Horus seems to include constellations—that is, groups of stars rising—but not single stars.

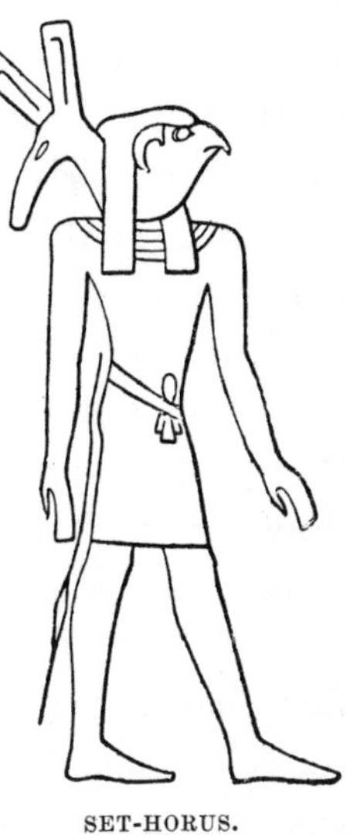

SET-HORUS.

Since the northern constellations were symbolised by the name of Set, the god of darkness, we should take Set-Horus to mean that the stars in the Dragon were rising at sunrise. This may explain the meaning of a remarkable figure which has set Egyptologists thinking a great deal. It is the combination of Horus and Set—a body of Horus with two heads, those of the hawk and jackal.

Now then for the myth. The reason why Naville went to the temple of Edfû for his facts is that in the later-time temples—and this is one of them—the inscriptions on the walls have chiefly to do with myth and ritual, whereas in the period covered by the earlier dynasties the temple inscriptions related chiefly to the doings of the kings. When we come to read the story which Naville brings before us, it looks as though the greatest antiquity must be conceded to it from the fact that the god Horus—the rising sun—is accompanied by the Hor-shesu, the followers or worshippers of Horus. These people are almost prehistoric, even in Egyptian history. De Rougé says of them, as I have previously pointed out, *C'est le type de l'antiquité la plus reculée.* They represent, possibly, the old sun-worshippers at a time when as yet there was no temple of the sun. Now, in this famous myth of Horus, Horus, accompanied and aided by the Hor-shesu, does battle with Typhon, the god of darkness, who had killed his father Osiris, and Horus avenges his father in the manner indicated in the various inscriptions and illustrative drawings given in the temple of Edfû. How does he do it? We find that in this conflict to revenge his father Osiris, he is represented in a boat killing a hippopotamus with ten darts, the beast being ultimately cut up into eight pieces. In some drawings it is a hippopotamus that he is slaying; in others, possibly for some totemic reason, a crocodile has been selected, but we can only see that it has been a crocodile by the fact that a little piece of the tail remains. Doubtless the reference had been found objectionable by some crocodile-worshipping people.

In very many inscriptions the constellation which, as I have stated, represents the hippopotamus, is really represented as a crocodile, or as a crocodile resting on the shoulders of a hippopotamus, so that there is no doubt that the crocodile and the hippopotamus were variants; and we can quite understand, further, that the hippopotamus must have been brought into Egypt by a tribe with that totem, who must have come from a very long way up the Nile, since the hippopotamus was never indigenous in the lower reaches of the river; so that we have in the myth to do with a hippopotamus-worshipping tribe, which, for that reason, probably came from a region very far to the south. There is evidence of local tribes in Egypt among which the crocodile was sacred.

The astronomical explanation of this myth is, I think, very clear. The inscriptions relating to one of the very earliest

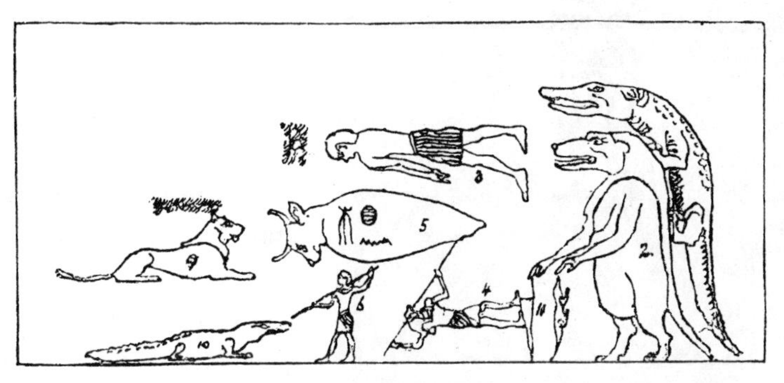

ILLUSTRATION FROM A THEBAN TOMB, SHOWING THE ASSOCIATION OF THE CROCODILE AND HIPPOPOTAMUS, AND HORUS SLAYING THE CROCODILE, AND THE CONSTELLATION OF THE THIGH.

of the illustrations refers to Horus, "the great god, the light of the heavens, the lord of Edfû, *the bright ray which appears on the horizon.*" The myth, therefore, I take it, simply means that *the rising sun destroys the circumpolar stars.* These stars are represented in the earliest forms of the myth either by the crocodile or the hippopotamus; of course they disappeared (or were killed) at sunrise. Horus, the bright ray on the horizon, is victorious by destroying the crocodile and the hippopotamus, which represent the powers of darkness.

This is a general statement. I should not make it if I could not go a little further. There is an astronomical test of its validity, to which I must call attention. The effect of precession is extremely striking on the constellations near the pole, for the reason that the pole is constantly changing, and the changes in the apparent position of the stars there soon become very obvious. The stars in Draco were circumpolar, and could, therefore, have been destroyed (or rendered invisible), as the hippopotami were destroyed in the myth by the rising sun, about 5000 years B.C.; and be it noted that at that time there was only one star in the Great Bear (or the Thigh) which was circumpolar. But at 2000 years B.C. the stars in Ursa Major were the circumpolar ones, and the chief stars in the constellation Draco, which formed the ancient constellation of the Hippopotamus, rose and set; so that, if there is anything at all in the explanation of the myth which I have given, and if there is anything at all in the idea that the myth is very ancient and refers to the time when the constellation of the Hippopotamus was really circumpolar—a time 7000 years ago—we ought to find that as the myth existed in more recent times, we should no longer be dealing with Draco or

HORUS AND CROCODILES.

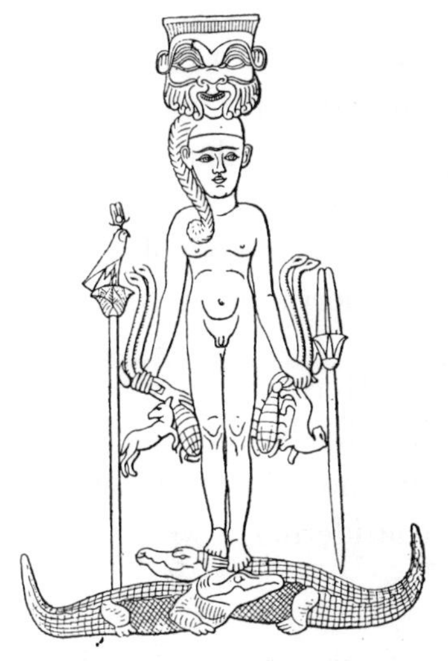
HORUS AND CROCODILES.

the Hippopotamus, because Draco was no longer circumpolar.

As a matter of fact, in later times we get Horus destroying no longer the Hippopotamus or the Crocodile, but *the Thigh of Set;* and, as I have said, 2000 years B.C. the Thigh occupied exactly the same position in the heavens with regard to the pole as the Hippopotamus or the Crocodile did 3000 years before.

Thus, I think, we may claim that this myth is astronomical from top to bottom; it is as old as, and probably rather older than, Naville thought, because it must certainly have originated in a period somewhere about 5000 years B.C, otherwise the constellation of the Hippopotamus would not have figured in it.

PTAH AND CROCODILES.

The various illustrations of Horus on the crocodiles are a reference to the myth we have just discussed.

It is easy to understand that if the myth were astronomical in origin there was no reason why it should be limited to Horus representing the rising sun; we accordingly find it extended to the god Ptah.

But although I hold that the *astronomical* meaning of the myth is that the rising sun kills the circumpolar stars, I do not think that is the last word. A conflict is suggested between a people who worship the rising sun and another who worship the circumpolar stars. I shall show in the sequel that there is an astronomical suggestion of the existence of two such distinct races, and that the companions of the sun-god of Edfû must probably be distinguished from the northern Hor-shesu.

Here we may conclude our reference to the stars which, in the latitude of Egypt, do not rise and set—or, rather, did not rise and set at the epochs of time we have been considering.

CHAPTER XV.

TEMPLES DIRECTED TO THE STARS.

I HAVE now to pass from the circumpolar stars to those which both rise and set. The difference between the two groups—those that do not rise and set and those which do—was fully recognised by the Egyptians, and many references are made to the fact in the inscriptions.

In a previous chapter I have given reasons to show that some of the earliest solar temples in Egypt were not oriented to the solstice.

The temple of Amen-Rā at Karnak, however, and others elsewhere were built in such a manner that at sunset at the summer solstice—that is, on the longest day in the year—the sunlight entered the temple and penetrated along the axis to the sanctuary. I also pointed out that a temple oriented in this manner truly to a solstice was a scientific instrument of very high precision, as by it the length of the year could be determined with the greatest possible accuracy, provided only that the observations were continued through a sufficient period of time.

All the temples in Egypt, however, are not oriented in such a way that the sunlight can enter them at this or any other time of the year. They are not therefore solar temples, and they cannot have had this use. The critical amplitude for a temple built at Thebes so that sunlight can enter it at sunrise or sunset is about 26° north and south of east and west, so that any temples facing more northerly or southerly are precluded from having the sunlight enter them at any time in the year.

It is imperative to be perfectly definite and clear on the question of the amplitudes above 26° at Thebes. I repeat, therefore, that any amplitude within 26° means that up to that point the sun at sunrise or sunset could be observed some day or days of the year—once only in the year if the amplitude is exactly at the maximum, twice if the maximum is not reached. But in the case of these temples with greater amplitudes than 26°, it is quite clear that they can have had nothing to do with the sun.

This being so, we have the problem presented to us whether or not temples were built so that starlight might fall along their axes in exactly the same way that the sunlight could fall along the axes of the solar temples when the sun was rising in the morning or setting in the evening.

It is abundantly clear that temples with a greater amplitude than 26° were oriented to stars if they were oriented at all by astronomical considerations. How can this question be studied? What means of investigation are at our disposal?

Suppose that the movements of the stars are absolutely regular; that there is no change from year to year, from century to century, from æon to æon; then, of course, the question as to whether or not these temples were pointed to a star, at rising or setting, would be easily and sufficiently settled by going to see; because if the stars did not change their apparent places in the heavens—accurately speaking, their declinations—and, therefore, the amplitudes at which they appear to rise and set, then, of course, a temple consecrated to Sirius ten thousand years ago would view the rising or setting of Sirius now as it did then.

But, as a matter of fact, astronomy tells us, as we have seen, that the apparent positions of the stars are liable to change. The change is much greater in the case of the stars than it is in the case of the sun, referred to in Chapters VI. and XI.; but still we have seen that the latter is one which has to be reckoned with the moment it becomes a question of inquiry into any time far removed from the present.

Hence, although in the case of the sun, there is, of course, no precessional movement, and although a temple once oriented to the sun would remain so for a long time; still, after some thousands of years, the change in the obliquity of the ecliptic would produce a small change in the amplitude at which a solstice is observed.

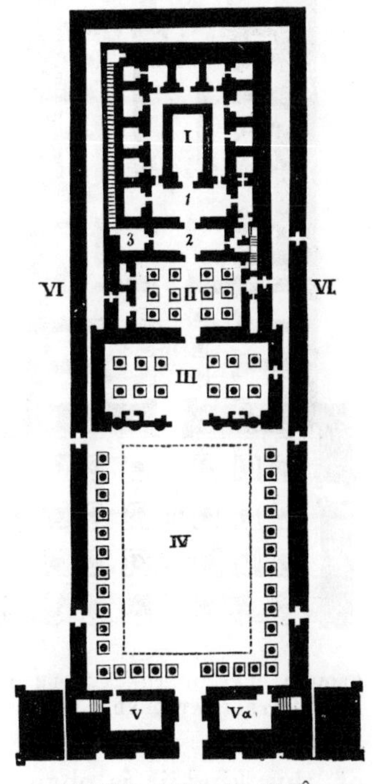

GROUND PLAN OF EDFÛ.

But while, in the case of the sun, we have to deal with a change of something like 1° in seven thousand years; we have to face in the case of the stars a maximum change of something like 47° in a period of thirteen thousand years. The change of declination must be accompanied by a change of amplitude, and therefore by a change in the direction of the temples.

Hence, when we get a temple of known date, with an amplitude which has been accurately measured, we can determine from that amplitude the exact declination of the body the temple was intended to observe, supposing, of course, that the temple was oriented upon any astronomical considerations at all. If the declination of the body turns out to be 23° 30′ or less, the temple may have been, in all probability was, a solar one; if the declination is greater it cannot have had anything to do with the sun directly.

This being so, it will be understood why in an inquiry

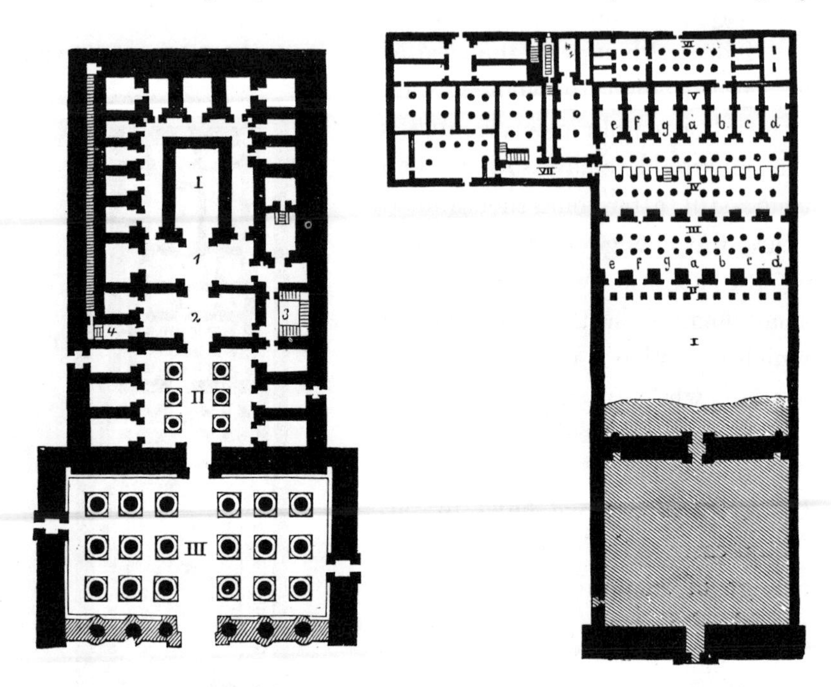

GROUND PLAN OF THE TEMPLE OF HATHOR AT DENDERAH.

PLAN OF THE TEMPLE OF SETI AT ABYDOS.

of this kind it is obviously desirable to begin with a region in which the number of temples is considerable. Such a condition we have in the region near Thebes; and the directions of the axes of the different temples—that is, the orientation of each of them, or, in other words, the amplitude of the direction in which each temple points—have all been tabulated. Chief among these we have the large temple of Karnak, showing that the amplitude of its orientation is 26° north of west, and the temple of Mut, showing that its orientation

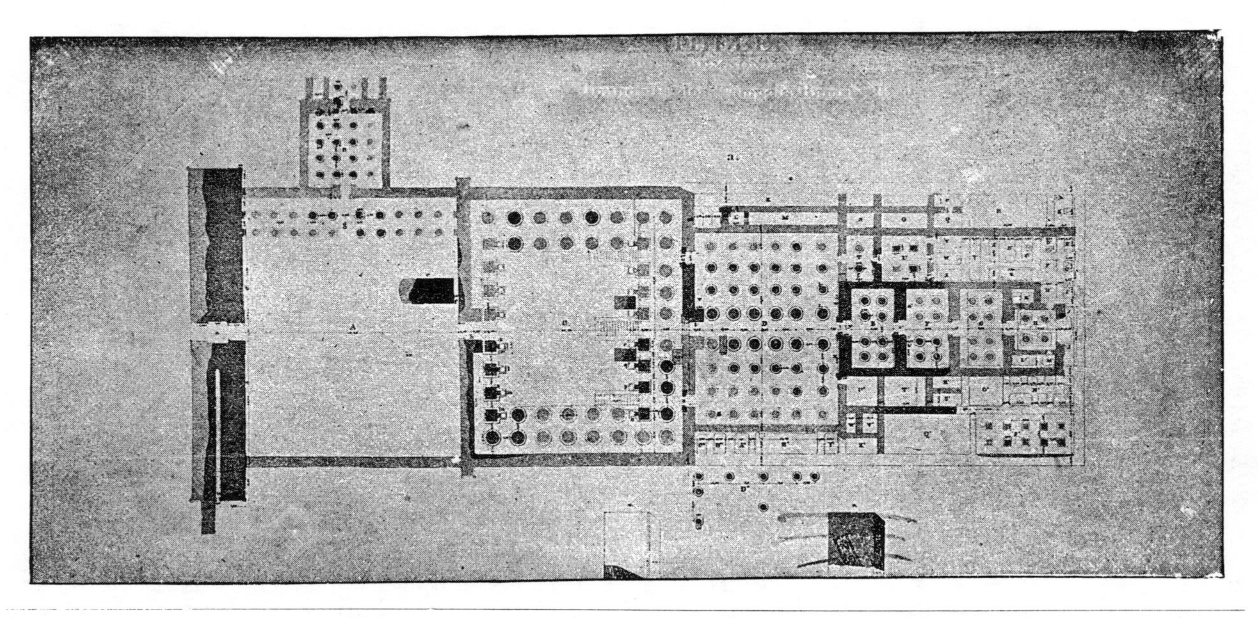

PLAN OF THE TEMPLE OF RAMESES II. IN THE MEMNONIA AT THEBES (FROM LEPSIUS), SHOWING THE PYLON AT THE OPEN END AND THE SANCTUARY AT THE CLOSED ONE.

is $72\frac{1}{2}°$ north of east. There is a temple at right angles to the temple at Karnak, and again another with an amplitude of 63° south of west, and so on.

It may be stated generally that at Karnak itself, not to go farther afield, there are two well-marked series of temples which cannot, for the reason given, be solar, since one series faces a few degrees from the north, and the other a few degrees from the south. There are similar temples scattered all along the Nile valley.

When we come to examine these non-solar temples, the first question is, Do they resemble the solar ones in construction? Are the horizontal telescope conditions retained? The evidence on this point is overwhelming. Take the Temple of Hathor at Denderah. It points very far away from the sun; the sun's light could never have enfiladed it; in many others pointing well to the north or south the axis extends from the exterior pylon to the Sanctuary or Naos, which is found always at the closed end of the temple; we have the same number of pylons, gradually getting narrower and narrower as we get to the Naos, and in some there is a gradual rise from the first exterior pylon to the part which represents the section of the Naos, so that a beam of horizontal light coming through the central door might enter it over the heads of the people flocking into the outer courts of the temple, and pass uninterruptedly into the Sanctuary.

In this way the Egyptians had, if they chose to use it, a most admirable arrangement for observing, with considerable accuracy, either the rising or the setting of any celestial body, whether it were sun or star, and especially the possibility of observing a *cosmical* rising, as the eye was shielded from the sun-rise light, and the place of rising was completely indicated.

In these, as at Karnak, we have a collimating axis. We have the other end of the temple blocked; we have these various diaphragms or pylons, so that, practically, there is absolutely no question of principle of construction involved in this temple that was not involved in the great solar temple of Amen-Rā itself.

We made out that in the case of the temples devoted to sun-worship and to the determination of the length of the year, there was very good reason why all these attempts should be made to cut off the light, by diaphragms and stone ceilings, because, among other things, one wanted to find the precise point occupied by the sunbeam on the two or three days near the winter or summer solstice in order to determine the exact moment of the solstice.

But if a temple is not intended to observe the sun, why these diaphragms? Why keep the astronomer, or the priest, so much in the dark? There is a very good reason indeed.

From the account given by Herodotus of the ceremonials and mysteries connected with the temple of Tyre, it is suggested that the priests used starlight at night for some of their operations, very much in the same way as they might have used sunlight during the day. According to Herodotus, in the temple in question there were two pillars—the one of pure gold, and the other of an emerald stone of such size as to shine by night. Now, there can be little doubt that in the darkened sanctuary of an Egyptian temple the light of α Lyræ, one of the brightest stars in the northern heavens, rising in the clear air of Egypt, would be quite strong enough to throw into an apparent glow such highly-reflecting surfaces as those to which Herodotus refers.

[1] Herodotus II., 44. (I am indebted to my friend Prof. Robertson Smith for this reference.)

Supposing such a ceremonial as this, the less the worshippers —who, reasoning from the analogy of the ceremonial termed the manifestation of Rā,[1] would stand facing the sanctuary, with their backs to the chief door of the temple—knew about the question of a bright star which might probably produce the mystery, the better.

Again, the truer the orientation of the temple to the star, and the greater the darkness the priest was kept in, the sooner would he catch the star quivering in the light of either early or late dawn.

In the first place, the diaphragms would indicate the true line that he had to watch; he would not have to *search* for the star which he expected; and obviously the more he was kept in the dark the sooner could he see the star.

Is there any additional line of evidence beyond the structural conditions of the temples that the Egyptians used these temples to observe the stars? Here a very interesting question comes in: a temple built at one period to observe a star could not go on for ever serving its purpose, for the reason that the declination of the star must change, as we have seen, by precession. Therefore a temple built with a particular amplitude to observe a particular star at one period would be useless later on.

We have here possibly a means of testing whether or not any of these temples were used to observe the stars. In those very early days, 3000 or 4000 years B.C., we must assume that the people who observed the stars had not the slightest idea of these possible precessional changes; they imagined that they were just as safe in directing a temple to a star as they were in directing a temple to the sun. But with a star changing its declination in an average way, the *same* temple

[1] See *ante*, p. 111.

could not be used to observe the *same* star for more than 200 or 300 years; so that at the end of that time, if they still wished to observe that particular star, they must either change the axis of the old temple, or build a new one. I have mentioned an average time as the change of the star's declination is involved.

Now this change of direction is one of the most striking things which have been observed for years past in Egyptian temples.

As a matter of fact, we find that the axes of the temples have been changed, and have been freely changed; that there has been a great deal of work done on many of the temples which are not oriented to the sun, in order to give them a twist.

Once a solar temple, a solar temple for thousands of years; once a star temple, only *that* star temple for something like 300 years, so that the conditions were entirely changed.

We get cases in which the axis of a temple has had its direction changed, and others in which, where it has been difficult or impossible to make the change in a temple, the change of amplitude has been met by putting up a new temple altogether. We are justified in considering such temples as a series in which, instead of changing the orientation of a pre-existing temple, a new temple has been built to meet the new condition of things. That, I think, is a suggestion which we are justified in making to Egyptologists on astronomical grounds.

For an instance, I may refer to the well-known temple at Medînet-Habû. We have there two temples side by side—a large temple, which was built later, with its systems of pylons and sanctuaries; a smaller temple, with outside courts, and, again, a sanctuary built much earlier. The direction of these two temples is very different; there is a difference of several degrees. It is very difficult indeed to understand why these two structures should have been built in that way if there were not some good reason for it. The best hitherto found is the supposed symmetrophobia of the Egyptians.

We find the same thing in Greece. There is the old Parthenon, a building which may have been standing at the time of the Trojan war, and the new Parthenon, with an outer court very like the Egyptian temples, but with its sanctuary more nearly in the centre of the building. It was by the difference of direction of these two temples at Athens that my attention was called to the subject.

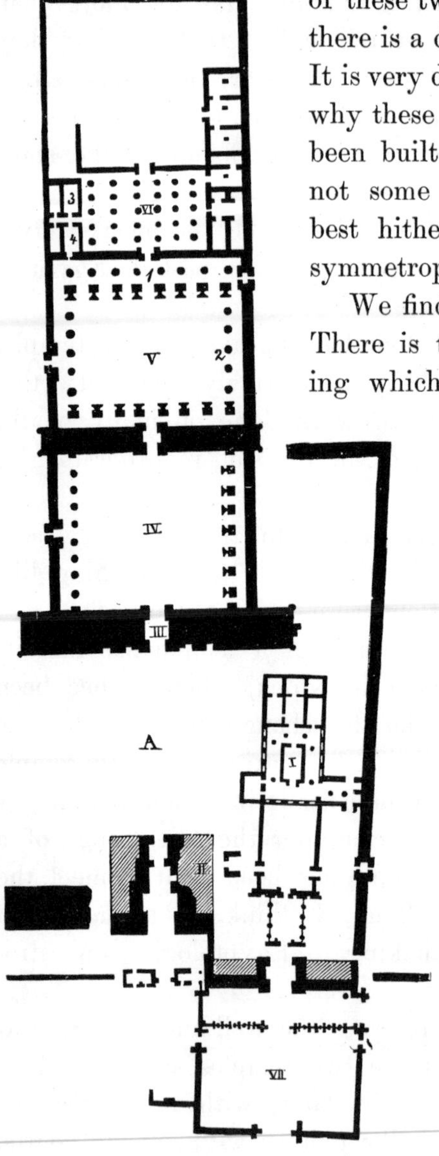

PLAN OF TWO TEMPLES AT MEDÎNET-HABÛ.

If we study the orientation of these, we find that, like those at Medînet-Habû, they are not parallel; there is a difference of orientation. This method of coping with the changes of amplitude of the star apparently represents that

THE BENT AXIS OF THE TEMPLE OF LUXOR, LOOKING TOWARD THE SANCTUARY.

adopted where there has been ample space to build another temple by the side of the old one when the star could no longer be seen from end to end of the old one. But another way was found where the space was more circumscribed, and that is well represented by the temple at Luxor, in which the addition is made *end on.* The suggestion is that, after the temple at Luxor had been built a certain number of years, the amplitude of the star had got a little out of the initial line, and the direction was changed at the time when it was determined to make the temple more beautiful and to amplify it by adding an outer court. There is another outer court and another very considerable change. There are four well-marked deviations.

CHAPTER XVI.

FURTHER INQUIRIES WITH REGARD TO THE STELLAR TEMPLES.

In the preceding chapter I discussed the suggestion, quite independently of any records the Egyptians may have left on the subject, that certain of the temples were oriented to stars; and I applied one test, that, namely, of the change of direction which was imperative if stars were observed for any lengthened period. In such an inquiry we must proceed with great caution.

We cannot make a statement regarding every particular temple with absolute certainty, for the reason that in the case of most of the temples the best Egyptologists cannot give us the most precious piece of information which we require from the astronomical point of view—that is, the date of the *foundation* of the temple. If in the case of these temples it were absolutely certain that each temple was built at a certain time with a certain orientation, we could tell at once whether or not that temple was pointed to any particular star.

In the absence of this precise information a general attack on the question has been necessary. The method adopted in the search has been as follows:—

(1) To tabulate the orientations of some of the chief temples described by the French Commission, by Lepsius and others.

Several interesting facts were soon revealed by this tabulation.

The first point that I have to note is that, in the case of some of these temples, we get the same, or nearly the same, amplitudes in different localities. To show this clearly it will be convenient to compare together the chief temples near Karnak and those having the same amplitudes elsewhere. We can do this by laying down along a circle the different amplitudes to which these various temples point. To begin with and to make the story complete, I draw attention to the temples which we have already discussed with an amplitude of 27°, or 26°, at Thebes, Karnak, and elsewhere. These, of course, are solar temples. Next we have non-solar amplitudes at Karnak and Thebes, associated with temples having the same amplitude at Denderah, Annu, and other places.

Another point is that we have the majority of the non-solar temples removed just as far as they can be in amplitude from the solar ones, for the reason that they are as nearly as possible *at right angles* to them, so that if the sun were observed in one temple and a star in the other, there would be a difference of 90° between the position of the sun and the position of the star at that moment. This would, of course, apply also to two stars. Sometimes this rectangular arrangement is in the same temple, as at Karnak, sometimes in an adjacent one, as at Denderah.

If we study Denderah we find that we have there a large temple enclosed in a square *temenos* wall, the sides of which are parallel to the sides of the temple; and also a little temple at right angles to the principal one.

It is hardly fair to say that a rectangular arrangement, repeated in different localities, is accidental; it is one which is used to some extent in our modern observatories.

The perpetual recurrence of these rectangular temples shows, I think, that there was some definite view in the minds of those who built all the pairs of temples which are thus related to each other; what that view was I shall endeavour to discuss in the sequel.

A third circumstance is that, when we get some temples pointing a certain number of degrees south of east, we get other temples pointing the same number of degrees south of west, so that some temples may have been used to observe risings and others settings of stars in the same declination. It is then natural, of course, to suggest that these temples were arranged to observe the rising and setting of the same stars; but further inquiry has shown that there are mythological objections to this explanation.

Finally, we have temples with the same amplitudes high north and high south, in different places—temples which could not have been built with reference to the sun; just as we have at different places temples with the same amplitudes which *could* have been used for solar purposes.

(2) To extend and check some of these observations with special reference to my new point of view in Egypt itself.

In connection with the possible astronomical uses of these temples, I find that when one of the temples has been built, the horizon has always been very carefully left open; there has always been a possibility of vision along the collimating axis prolonged. Lines of sphinxes have been broken to ensure this;[1] at Medînet-Habû, on the opposite side of the river to Karnak, we have outside this great temple a model of a Syrian fort. If we prolong the line of the temple from the middle of the Naos through the systems of pylons, we find that in the model of the fort an opening was left, so that the vision from the sanctuary of the temple was left absolutely free to command the horizon.

It may be said that that cannot be true of Karnak, because

[1] For instance, in the line of sphinxes in front of temple X, shown in the folding plate inserted in Chap. XVIII., the line was left incomplete to preserve the fair-way of the ruined temple north of Y outside the *temenos* wall.

we see on the general plan that one of the temples, with an azimuth of 72½° N., had its collimating axis blocked by numerous buildings. That is true; but when one comes to examine into the date of these buildings, as I propose to do in a subsequent chapter, it is found that they are all very late; whereas there is evidence that the temple in question was one of the first, if not the very first, of the temples built at Thebes.

(3) To determine the declinations to which the various amplitudes correspond. In this direction I have made use of the German Catalogue of star places from 1800 A.D. to 2000 B.C., the places for dates beyond this, and for southern stars, having been calculated chiefly by my son, Mr. W. J. S. Lockyer, B.A.

Some places for Sirius and Canopus have been obligingly placed at my disposal by Mr. Hind, and approximate values obtained by the use of a precessional globe constructed for me by Mr. Newton. This globe differs considerably from that previously contrived by M. Biot, about which I was ignorant when I began the work, and enables right ascensions and declinations, but especially the latter, to be determined with a fair amount of accuracy for forty-eight equidistant points occupied by the pole of the equator round the pole of the ecliptic (assumed to be fixed) in the precessional revolution.

Some simple astronomical considerations may here come to our help. If the north polar distance of a star is increasing—that is, if a star is increasing its distance from the north pole—its declination if north or south will be decreased or increased respectively, and the orientation of the temple would be gradually becoming more and more parallel to an E. and W. line; if the declination north or south of the star be increasing, then the orientation of the temple would have to be likewise increased. The change in the orientation, therefore, gives us information towards determining in which quarter of the heavens each particular star might have been.

(4) In cases where the date of the foundation of a temple dedicated to a particular divinity has been thoroughly known, there was no difficulty in finding the star the declination of which at the time would give the amplitude; and, in the case of series of temples dedicated to the same divinity, an additional check was afforded if the changes of amplitude from the latest to the newest temple agreed with the changes of the declinations of the same star.

(5) Having the declinations of the stars thus determined for certain epochs, I have next plotted them on curves, showing the amplitude for any year up to 5000 B.C. at Thebes for a true horizon and when the horizon is raised 1° or 2° by hills or mist; and, finally, a table has been prepared showing the declination proper to the amplitude of each of the chief temples when the needful information was available.

Although, however, these matters can be discussed in a way that will indicate that the inquiry is raised, I do not wish for one moment to speak of it as being settled, because the observations which have been made already in Egypt with regard to the orientation of these temples have not been made from such a very special point of view; and, further, considerable alteration in the amplitude would be made by the presence of even a low range of hills miles away in the case of stars rising or setting not many degrees from the north or south. No one would care to make the assertion with absolute definiteness until it was known whether or not the horizon in each case was interfered with by hills or any intervening objects—was or was not one, in fact, which might be regarded as a sea horizon from the point of observation; if there were impediments, the angular height of them must, of course, be exactly known; but this information is almost entirely lacking.

Now, however, that the question has been raised by observations of the temples themselves, it becomes interesting to ask of the inscriptions if there are records that these temples were directed to stars?

It will be seen in the next chapter that the inscriptions give out no uncertain sound on this point.

CHAPTER XVII.

THE BUILDING INSCRIPTIONS.

NUMEROUS references to the ceremonial of laying the foundation-stones of temples exist, and we learn from the works of Chabas, Brugsch, Dümichen,[1] and others, that the foundation of an Egyptian temple was associated with a series of ceremonies which are repeatedly described with a minuteness which, as Nissen has pointed out,[2] is painfully wanting in the case of Greece and Rome. Amongst these ceremonies, one especially refers to the fixing of the temple-axis; it is called, technically, "the stretching of the cord," and is not only illustrated by inscriptions on the walls of the temples of Karnak, Denderah, and Edfû—to mention the best-known cases—but is referred to elsewhere.

Another part of the ceremony consisted in the king proceeding to the site where the temple was to be built, accompanied mythically by the goddess Sesheta, who is styled "the mistress of the laying of the foundation-stone."

Each was armed with a stake. The two stakes were connected by a cord. Next the cord was aligned towards the sun or star, as the case might be; when the alignment was perfect the two stakes were driven into the ground by means of a wooden mallet; there was no difference of procedure in the case of temples directed to the sun. One boundary wall parallel to the main axis of the temple was built along the line marked out by this stretched cord.

[1] "Baugeschichte des Dendera-Tempels," 1877.

[2] "Rheinisches Museum für Philologie," 1885, p. 39.

If the moment of sun- or star-rise or -set were chosen, as we have every reason to believe was the case seeing that all the early observations were made on the horizon, it is obvious that the light from the body towards which the temple was

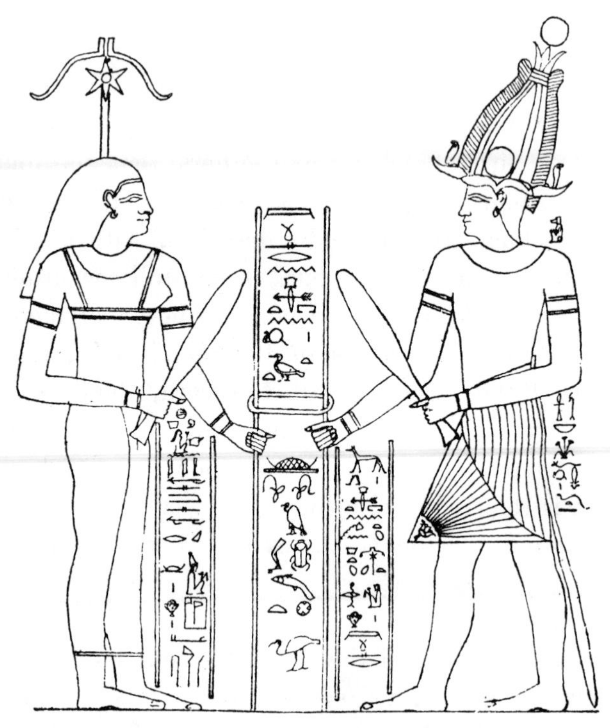

THE LAYING OF THE FOUNDATION-STONE CEREMONIAL.

aligned would penetrate the axis of the temple thus built from one end to the other in the original direction of the cord.

We learn from Chabas that the Egyptian word which expresses the idea of founding or laying the foundation-stone of a temple is *Senti*—a word which still exists in Coptic. But in the old language another word, *Put-ser*, which no longer remains in Coptic, has been traced. It has been established that *put* means "to stretch," and *ser* means "cord;" so that that part of the ceremonial which consisted in stretching a cord in the direction of a star was considered of so great an importance that it gave its name to the whole ceremonial.

I will next refer to some of the inscriptions; one, dating from the last half of the third thousand B.C., occurs in the document describing the building of the temple of Annu (Heliopolis). We read:—"Arose the king, attired in his necklace and the feather crown; all the world followed him, and the majesty of Amenemhāt [first king of the Twelfth dynasty]. The Kher-heb read the sacred text during the stretching of the measuring-cord and the laying of the foundation-stone on the piece of ground selected for this temple. Then withdrew His Majesty Amenemhāt; and King Usertsen [son and co-regent] wrote it down before the people."

Nissen, from whom (*loc. cit.*) I quote the above, adds:—"On account of the stretching of the measuring-cord, the Egyptian engineers were called by the Greeks ἀρπεδονάπται, whose art Democritus boasts of having acquired."

We next turn to Abydos, possibly one of the oldest temple-fields in Egypt. There is an inscription relating to the rebuilding of one of them in the time of Seti I. (about 1380 B.C.). In this the goddess Sesheta addresses the king as follows:—"The hammer in my hand was of gold, as I struck the peg with it, and thou wast with me in thy capacity of Harpedonapt. Thy hand held the spade during the fixing of its [the temple's] four corners with accuracy by the four supports of heaven." On the pictures the king appears with the Osiris crown, opposite the goddess. Both hold in their right hand a club, and with it they each hammer a long peg into the ground. Round the two pegs runs a rope, which is stretched tight, the ends being tied together.

In two cases the star used for the alignment is actually named. Of these I will take, first, the record of the ceremony used in the building of the temple of Hathor at Denderah.

THE ALIGNMENT OF THE TEMPLE OF HATHOR AT DENDERAH.

The inscriptions state that the king while stretching the cord had his glance directed to the *āk* of the constellation of the *Thigh*—the old name of the constellation which we now recognise as the Great Bear—and on this line was built the new temple, "as had been done there before."

The actual inscription has been translated as follows:—"The living God, the magnificent son of Asti [a name of Thoth], nourished by the sublime goddess in the temple, the sovereign of the country, stretches the rope in joy. With his glance towards the *āk* [the middle?] of the Bull's Thigh constellation, he establishes the temple-house of the mistress of Denderah, as took place there before." At another place the king says: "Looking to the sky at the course of the rising stars [and] recognising the *āk* of the Bull's Thigh constellation, I establish the corners of the temple of Her Majesty."

Here, then, we have more than evidence of the stretching of a cord towards a star; an actual constellation is named, and it may be easily imagined that in connection with this many interesting questions arise of the utmost importance to the subject we are considering.

Dümichen, in his references to this passage, discusses the meaning of the word *āk* in relation to some Theban grave-inscriptions, in which it is suggested that *āk* is used to represent the middle course of a star, or, astronomically speaking, its culminating point as it passes the meridian. But such a meaning as this will never do in this connection; for if a cord was stretched towards a star on the meridian it would lie north and south, and therefore the temple would be built north and south. But this is by no means the orientation of the temple—a point to which I shall return presently.

But it may be suggested that the word *āk*, used in relation to the king's observation, more probably referred to the brightest star, Dubhe (α Ursæ Majoris) in the asterism, or the "middle point" of the constellation, which would be about represented by the star δ, which lies nearly in the centre of the modern constellation of the Great Bear, supposing, indeed, that the same stars were included in the old constellation. On this point we unfortunately have no definite knowledge, as the Thigh is so variously represented; sometimes there is a hind-quarter, represented evidently by the well-known seven stars; at others the body of a cow (with horns and disk) is attached.

However this may be, without such a reference to some particular part of the constellation it is obvious that the stretched cord may have had a most indeterminate direction.

In order to leave no stone unturned in attempting to explain this description—supposing it to represent an undoubted fact of observation—we may consider another possible interpretation of the word *āk*. The amplitude of the temple being 71½° N. of E., shows conclusively that we cannot be dealing with the meridian, but may we be dealing with the most eastern elongation of the star in its journey round the Pole?

I have inquired into this matter for the time of the last building of the temple in the time of the Ptolemies, and find that the amplitude of the temple, instead of being 71½°, would have been about 70°. It seems probable, then, that this interpretation will not hold, and it may be further stated that, in the case of a star at a considerable angle above the horizon, the stretching of a cord in the building ceremonial—the "Ausspannung der Strickes," as the words *put-ser* are translated by Dümichen—would really have been no stretching of the cord at all; for the star being many degrees above the horizon, another method must have been employed, and in all probability would have been distinctly referred to in the careful statements of the ceremonies which exist. I think, then, that we are perhaps justified in dismissing this possible explanation, especially as rising stars are referred to.

We now come to considerations of a different order. The inscription which we have quoted is put into the mouth of the Emperor Augustus, though he never was at Denderah.

This suggests that the temple built in the time of Augustus carried forward the account of the old foundation. There is evidence of this. The constellation of the Thigh neither rose nor set in the time of Augustus—it was circumpolar. The same statement may be made regarding the restoration in the time of Thothmes III. So we are driven to the conclusion that if we regard the inscription as true, it must refer to a time preceding the reign of Thothmes. I shall return to this subject in a subsequent chapter.

THE ALIGNMENT OF THE TEMPLE OF EDFÛ.

A reference to the same constellation (the Thigh) is also made in the account of the ceremonial used at the laying of the foundation-stone of the temple at Edfû. The king's glance was directed—in the case of the building of that temple—to the *Thigh*, but no precise reference to any star or to any point *āk* is given.

As before, I give the full translation of the inscription,[1] remarking that the last restoration was made B.C. 237—57.

[1] Quoted from Nissen, *op. cit.*

The king is represented as speaking thus:—"I have grasped the wooden peg and the handle of the club; I hold the rope with Sesheta; my glance follows the course of the stars; my eye is on Mesχet [that is, the 'Bull's Thigh constellation,' or Great Bear]; (mine is the part of time of the number of the hour-clock); I establish the corners of thy house of God." And in another place:—"I have grasped the wooden peg; I hold the handle of the club; I grasp the cord with Sesheta; I cast my face towards the course of the rising constellations; I let my glance enter the constellation of the Great Bear (the part of my time stands in the place of his hour-clock); I establish the four corners of thy temple." The translation is Brugsch's. The phrases in parentheses are interpreted differently by Dümichen, who translates them:—"Standing as divider of time by his measuring instrument," or "representing the divider of time (*i.e.* the god Thoth) at his measuring instrument." The word *merech* or *merchet*, in which Brugsch suspects hour- or water-clock, does not occur elsewhere.

In this case, seeing that the temple lies with its axis very nearly north and south, as I determined by my own (magnetic) observations, the stretching of the cord was certainly in or very near the meridian; and it may be remarked that in the Naos there is an opening in the roof, over the side of the second or third door from the sanctuary, and inclined at an angle of 40° (unlike any other opening that I have seen in the roof of any Egyptian temple), which may have been used to observe the transit of some particular star. The angle I was not able to determine with absolute accuracy, as the vertical circle of the theodolite I had with me was out of adjustment.

Taking the latitude of Edfû as 25°, and assuming the angle of 40° to be not far from the truth, the North Polar distance of the star observed would be 15°.

Within a degree or so—and this is as near as we can get till more accurate observations have been made on the spot—this satisfies Dubhe, the chief star in the Great Bear in the time of the Ptolemies. Supposing the temple was originally oriented to Dubhe, its amplitude, $86\frac{1}{2}°$ S. of W., gives us the date 3900 B.C. I shall show, however, that it is more probable that the temple was oriented on some southern star.

I may here remark that, so far as I know, Edfû is the temple in Egypt nearest the meridian. If, therefore, it were used as, on my theory, all other temples were, it could only have picked up the light from each of the southerly stars, as by the precessional movements they were brought into visibility very near the southern horizon.

In this respect, then, it is truly a temple of Horus, in relation to the southern stars—the southern eyes of Horus. But it was not a sun-temple in the sense that Karnak was one; and if ceremonies were performed for which light was required, perhaps the apparatus referred to by the writer Dupuis[1] was utilised. He mentions that in a temple at Heliopolis—whether a solar temple or not is not stated—the temple was flooded all day long with sunlight by means of a mirror. I do not know the authorities on which Dupuis founds his statement, but I have no doubt that it is amply justified, for the reason that doubtless all the inscriptions in the deepest tombs were made by means of reflected sunlight, for in all freshly-opened tombs there are no traces whatever of any kind of combustion having taken place, even in the innermost recesses. So strikingly evident is this that my friend M. Bouriant, while we were discussing this matter at Thebes, laughingly suggested the possibility that the electric light was known to the ancient Egyptians.

[1] "Origine des Cultes," vol. i., p. 450.

With a system of fixed mirrors inside the galleries, whatever their length, and a movable mirror outside to follow the course of an Egyptian sun and reflect its beams inside, it would be possible to keep up a constant illumination in any part of the galleries, however remote.

Dupuis quotes another statement that the greatest precautions were taken that the first rays of sunlight should enter a temple (of course, he means a solar temple).

But it is possible that there might have been another temple at right angles, facing nearly due east. In this case, the larger temple would have been named after the worship to which the smaller one was dedicated. If so, unlike the solar temples at Heliopolis, Abydos, and Thebes, the Edfû temple was sacred to the Equinoctial Sun, or, at all events, to the Sun very near an equinox.

CHAPTER XVIII.

THE STAR-TEMPLES AT KARNAK.

WHEN I began my studies of the Egyptian temples the building inscriptions referred to in the preceding chapter lay forgotten in the Egyptologist's archives. I purpose now to give some account of my work at Thebes, where I made a special study of the temples, because there is a very great number there, and many are in a fair state of preservation. These investigations convinced me that temples were oriented to stars before the inscriptions in question were known to me, although the whole temple field is so crowded with temples, each apparently blocking up the fair-way of the other, that it seems well-nigh impossible that any such process as that described in the last chapter could have been applied.

This difficulty will be gathered from pages ii and iii that show a reproduction of Lepsius's general maps of the temple region of Karnak, showing his reference letters and also the *true* north and the orientation of the chief temples. We have already dealt with the solar temple of Amen-Ra.

We find, beginning at the south, a large temple with a long line of sphinxes, the temple of Mut (X) facing the large temple of Amen-Ra (K). To the north of the latter is another temple system (A and B and C), also with an avenue of sphinxes. On the east side of K another temple (O) is only slightly indicated.

To the south of the large temple K is another one—that of Khons (T), also with its sphinxes. Connected with K are two other temples, L, nearly, and M, exactly, at right angles to it. There is also such a rectangular temple (Y) added to the temple of Mut. I also call attention to the temples V and W, chiefly to point out that when I went over the ground with M. Bouriant it seemed to us as if the temple V faced S.E. and not N.W. as indicated by Lepsius. Very few traces of the temple are left.

Since the labours of the French and Prussian Governments gave full records of Karnak a memoir on the temples has been published by Mariette, which gives us not only plans, but precious information relating to the periods at which, and the kings by whom, the various parts of the temples were constructed or modified. No doubt those which are still traceable form only a very small portion of those which once existed; but however that may be, I have now only to call attention to some among them.

I have previously shown that the magnificent work of Mariette has supplied us with building dates for the solar temple to which reference has been made; so that we have, with more or less accuracy, the sequence of the various parts of the completed building.

If we consider the plan without any reference to the building dates at all, the idea that the smaller temples were built for observations of stars seems to be entirely discountenanced. The temple L, for instance, instead of having a clear horizon, is blocked by the very solid wall (2) and its accompanying columns; the temple M, instead of having a clear horizon, is absolutely blocked by two of the line of pillars (1) very carefully built in front of it. But if we consult Mariette, *we find in both cases that the wall was built long after one temple, and the pillars were built long after the other.*

This result is satisfactory, inasmuch as it indicates that

a natural objection to the orientation hypothesis is invalid. But can we strengthen it by supporting Mariette's statement as to the dates?

Mariette states that the temple M was built by Rameses III., a king of the twentieth dynasty. With this datum, we consider the orientation of the temple. The problem is one of this kind:—Taking the Egyptologist's date for Rameses III. at 1200 B.C., and taking the amplitude of the temple as 63½° N. of E., was there, when that temple was built, any star opposite to it, any star to which it accurately pointed? We can translate the amplitude of that temple into the declination of a star, making a slight correction for the stated conditions of observation in Egypt, which would make the apparent amplitude less than the true one, because the star would appear to rise more to the south. In the absence of precise information, we are justified in taking the mean of the values referred to by Biot—that is, an apparent amplitude due to a stratum of haze 1½° high, especially as the temple looked away from the Nile.

Searching the astronomical tables, we find that there was a star visible along the temple axis. The star was γ Draconis.

So much for the temple M. We now proceed to the other one lettered L, the temple of Seti II.

The amplitude of temple L is 63° S. of W., and the date, according to Mariette, 1300 B.C. We find the declination, proceeding as before, and assuming hills 1½° high, to be 53½° S., and about that date the bright star Canopus set on the align ment of the temple.

It will hence be gathered that just as truly as the temple M seems to have been pointed to the northern star γ Draconis rising, the temple L was pointed to the southern star Canopus, setting.

But this is not all. There is another temple to which I have already directed attention—the temple of Khons (T of Lepsius), founded by Rameses III., though as it comes to us it is a Ptolemaic structure, it having been enlarged and restored by the Ptolemies. It is very nearly, but not quite, parallel to the temple of Seti II.

My measures and those of Lepsius give, approximately, amplitudes as under—

Temple of Seti (T)	63°	S. of W.
Temple of Khons (L)	62°	,,

Continuing, therefore, the same line of inquiry, and assuming that Mariette was right, and that the temple was really finally completed (and no doubt its axis revised) by the Ptolemies, and that they flourished about 200 B.C., we have the same problem. Was there a star towards which that temple could have been directed, and which could have been seen in that temple with its actual orientation?

Calculation shows that the change of amplitude of Canopus due to the precessional movement between 1300 B.C. and 200 B.C. is almost exactly 1°, the difference in the amplitude of the temples. We seem, then, to have in the temples L and T two temples directed to the same star at different times.

These statements must be taken as provisional only. To render them absolute, careful measurements must be made, on the spot, of the heights of the hills towards which the temples point.

Leaving this for the moment on one side, we get in this manner astronomical dates of the reigns of Seti II. and Rameses III. within a very few years of those given by the Egyptologists.

More than this, the application of this method entirely justifies

Mariette's view with regard to these more modern temples at Thebes, and shows that when they were built the outlook was clear, so that the building ceremonials referred to in the last chapter might have been performed.

I am next anxious to point out that not only is this so, but, accepting it, we can explain exactly why the walls and temples and columns were erected in the sequence which Mariette indicates. We not only know when they were built, but we can presently understand *why they were built.*

The first point to which I draw attention in this matter is the following:—Referring to the plan, we find that before the time of Rameses III. the temple of Seti II. was right out in the open. It thus represented just one of those external rectangular temples which have been found at Denderah and at very many other places in Egypt. It was one of the Egyptian ideas to have two temples at right angles to each other. That temple, then, stood alone. The next change seems to have been this: The star Canopus, the setting of which it was built to watch, was, through the precessional movement to which I have referred, no longer conveniently observed in that temple. To obviate this the temple T was built by Rameses III. with a change of amplitude equivalent to the actual precessional change of the star's declination, to carry on the observations.

Further, at the same time another temple (M) was built to observe γ Draconis. It is now easy to understand what the 21st—a Theban—dynasty did. Seti's temple (L) had been superseded; the temple M was a second rectangular temple outside the great temple of Karnak (K). They said to themselves: "We will make Karnak more beautiful, and we will extend it. We can now build walls in continuation of the old walls, and we can build still another pylon, because Seti's temple is no longer being used, the worship having been transferred to the temple of Rameses III. (Khons). By building the northern wall we prevent the use of temple M, sacred to our enemy Sutech."

I should add that the opening in the wall, in prolongation of the axis of temple M, is *not* directly opposite the temple M, but a little to the east; it was probably made later, possibly by the twenty-second dynasty, who were Set worshippers. Again, coming to the time of Taharqa, returning at the end of the exile of the priests of Amen in Nubia, the temple M was again thrown out of use. Pillars were built in front of it, right in the fair-way, affording an instance that when a temple was thrown out of use, not by the precessional movement of the star to which it had been directed, but by the partisans of another creed, the fact of its being no longer in operation was insured by something being built in front of it, to prevent observation of the stellar divinity no longer in vogue.

It may be added that long after the temple of Seti II. fell out of astronomical use, and was on that account blocked by the walls of the twenty-first dynasty, the Ptolemies built a new temple of Osiris, which, if built before, would have been in the fair-way of the temple of Seti. Thus, there is a reason for all the changes made at all the dates referred to by Mariette.

I think we find in this result of the inquiry a valuable corroboration of Mariette's conclusions, and another reason why we should not cease to admire his magnificent work.

So far I have only referred to the relatively modern parts of Karnak. I now pass to the more ancient ones, in which we ought to note the same laws holding good, if there be any value in the view we are discussing.

We find that some of the most important temples given by Lepsius and Mariette (B, X, and W) are just as effectively blocked by the mass of the temple of Amen-Rā as those we have already considered were by the walls of the twenty-first dynasty and Taharqa's columns; and, looking at the plan, it seems at first perfectly absurd to continue to hold for one moment the idea that these temples were built for observations of stars on the horizon.

The temple X (Mut) is blocked by the pylon marked 3, the temple B by the eastern end of the great temple, the temple W by the temple O.

Mariette here again comes to our rescue to a certain extent. He shows, as I have stated in Chapter XI., that in the beginning of things, certainly in the twelfth dynasty, possibly in the eleventh dynasty, and possibly even before that, only the central part, marked 4, of the solar temple existed, less as a temple than as a shrine, with nothing to the west of it and nothing to the east of it.

That being so, the temple B gets its fair-way to the south, and the temple of Mut (X) and the smaller temple (W) to the north.

Mariette in his two plates shows the growth of the temple of Amen-Rā in a most admirable way, from the central portion of the temple to which I have referred—that is, the small central court, which, he is careful to note, existed before Thothmes I.; how much before, he does not say. Afterwards, the pylons are added; then they are elaborated; then the sanctuary is thrown back to the eastward, and the temple O built, and B thereby blocked, and then thrown forward to the westward, thus blocking X and Z.

If there is anything in these considerations at all, it is suggested that all the temples to which I have referred were founded before these easterly and westerly extensions, of which Mariette gives us such ample evidence.

In a subsequent chapter it is suggested that this great lengthening of the original shrine of Amen-Rā was undertaken for the purpose of blocking temples X, Z, and W, all dedicated to Set. Thothmes III. and Taharqa had precisely the same objects in view, apparently.

Here, however, we meet a real difficulty. Mariette states that, so far as he has been able to find, the temple B, a temple of which the worship is Amen, and the temple X, in which the worship is Mut, were built by Amen-hetep III. If that were so, they would have been built blocked; none of the usual ceremonials could have been employed at their foundation. They could not have been used at all for astronomical purposes, because their horizons were blocked by these extensions of the temple of Amen-Rā.

Here I must refer specially to temple B. Its amplitude is, according to Lepsius, $63\frac{1}{2}°$ S. of W. I have already shown that the amplitudes of temples L (Khons) and T (Seti II.) are 62° and 63° S. of W., and that in the times of the Ptolemies and Seti II., each faced the star Canopus in turn. Hence the probability that we have three temples of nearly equal orientation sacred to the same divinity.

Temple.	Orientation.	Declination.	Date.
Khons ...	62 ...	$52\frac{1}{2}°$ ...	300 B.C.
Seti II. ...	63 ...	$53\frac{1}{2}°$ ...	1350 B.C.
B ...	$63\frac{1}{2}$ ...	54° ...	1800 B.C.

The statement is that the part of the temple of Amen-Rā, the building of which blocked B, was commenced by Thothmes III., whose date, according to Brugsch, is 1600 B.C., and continued by Amen-hetep III. (1500 B.C.). Unless, then, some other provision was made, the observations of Canopus were not continued until another shrine was built. We know that another shrine was built, that of Seti II., and that its orientation gives a date of 1350 B.C. It might have been commenced by Seti I. after the Khu-en-Aten troubles, and finished by Seti II.

One is therefore tempted to ask whether we have not here one of those crucial cases which Mariette himself contemplated, in which the true foundation is so far anterior to the last restoration or the last decoration, from which, for the most part, the archæologist gets his information, that one is absolutely misled by the restorations or decorations as to the true date of the original foundation of the shrine.[1]

If the archæologists are right in attributing the granite temple of Osiris (?), near the sphinx, to a date anterior to, or even contemporaneous with, the second pyramid, we have evidence that in the early dynasties the temple building in stone, and even in granite brought from Aswân, was as perfect in the matter of workmanship as in the eighteenth dynasty; and that it was not then the fashion to inscribe walls, but only statues and stelas. May it possibly be that the fashion in question came in, or reached its greatest development, during the eighteenth dynasty, and that on this account so many temples are ascribed to that period, whereas they were actually in existence before?

If the prior dynasties built no temples, why did they not

[1] On this point I am permitted by Professor Maspero to print the following extract from a letter I received from him:—"Tous les temples ptolémaïques et la plus grande partie des temples pharaoniques sont des *reconstructions*. Ce que vous avez observé de Dendérah, est vrai d'Esnéh, d'Ombos, d'Assouan, de Philæ, etc. Or, si les premiers constructeurs d'un temple—ou chez nous d'une église—peuvent choisir presque à leur gré l'emplacement, et par suite l'orientation, la plus convenable, il en est bien rarement de même des *reconstructeurs*. Les maisons accumulées autour du temple les gênaient, d'ailleurs les habitudes du culte et de la population étaient prises; on rebâtissait le temple—comme d'ordinaire chez nous on rebâtit l'église—sur la même orientation et sur les mêmes fondations. J'ai constaté le fait à Kom-Ombo, où les débris du temple décoré par Amenhotpou I. et Thoutmosis III. sont orientés exactement comme ceux du temple ptolémaïque actuel, bâti *sur* les ruines du précédent. Vous avez donc le droit de dire, non seulement pour Dendérah, mais pour beaucoup d'autres temples, qu'ils ont été reconstruits sur l'orientation du temple qu'ils remplaçaient, quand même cette orientation ne répondait plus à la réalité des choses."

do so? and if they did, where are they, if some of those *inscribed* by the eighteenth dynasty be not they?

In the absence of final archæological evidence—that is, admitting Mariette's own doubt as to the mere existence of inscriptions—are there any astronomical considerations which may possibly help us? Assuming that the temples were astronomically oriented, we have *one* registering for us the time elapsed since the original direction of the axis was laid down, in terms of the change in the obliquity of the ecliptic.

We have others registering time in like manner in terms of the change due to precession, if we can get any light as to the stars towards which the temples were oriented.

I have already dealt with the temple of Amen-Rā in Chapter XI., and we found a foundation date of 3700 B.C. for the original shrine, so far as the rough observations already available can be trusted. Assuming the accuracy of this determination, it is clear that we must look for stars with appropriate amplitudes between that date and say 2500 B.C.

Let us take the temple of Mut (X of Lepsius); its amplitude is 72½ N. of E. This was the amplitude of γ Draconis about 3500 B.C. This temple, then, bore the same relation to M as T did to L! We have two cases of two temples erected at different dates to the same star.

Although it has been convenient to begin with Thebes for the reasons given, the records concerning any one temple there are far more restricted than those which relate to some temples elsewhere; while the cult can only be determined in few instances. I propose, therefore, for the present to content myself with the above general considerations showing the first application of the method of investigation adopted, and to pass on to Denderah, where we are sure of the cult and where many particulars are given.

CHAPTER XIX.

THE PERSONIFICATION OF STARS—THE TEMPLE OF ISIS AT DENDERAH.

WE have now to pass from the building ceremonials and a general consideration of the temples at Karnak, to the

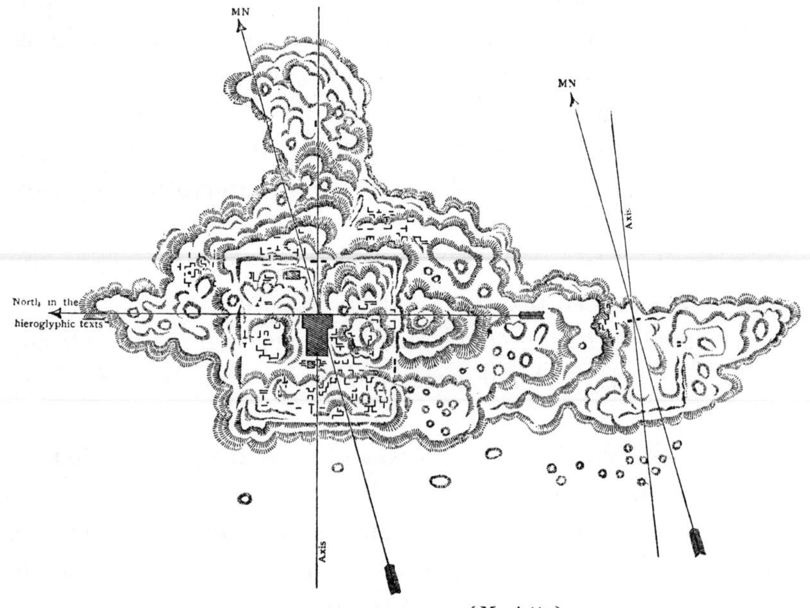

PLAN OF DENDERAH. (*Mariette.*)

worships to which the various temples were dedicated. And to do this we must face the problems of Egyptian mythology, so far as the names and origins of the various gods and goddesses are concerned.

There is ample evidence that each temple was sacred to

some god or goddess, although in many cases the name of the patron divinity has been lost.

Fortunately, at Denderah the patron divinities are well known, so it will be well to begin with the temples there. We find a general plan of Denderah among the magnificent drawings which we owe to the French expedition of 1798. This shows the wall round the temple-space containing the temple of Hathor, the great temple; and the smaller temple of Isis at right angles to it. We find, roughly, that the great temple points to the north-east; the smaller temple of Isis points to the south-east. A later plan has been published by Mariette in his work on Denderah.

These, then, are the main conditions of the temples at Denderah. But we can go a little more closely into them by referring to the map which accompanies Biot's memoir, to which I have previously referred. He gives the axis of the Hathor temple pointing, not merely to the north-east, but to 18° E. of N. Since the other temple lies at right angles to the great one, its direction, according to Biot, is 18° S. of E.

To show the uncertainty in these inquiries brought about by the absence of a proper survey, I may give the following later values:—

1. LEPSIUS, 1844—
   Magnetic azimuth of the axis ... ... N. 25° E.
   ,, amplitude ,, ,, ... 65° N. of E.
   Correction $8\frac{1}{2}°$
   ∴ Astronomical amplitude ... ... $73\frac{1}{2}°$ N. of E.
2. MARIETTE, 1870—
   Astronomical azimuth ... ... N. 15° E.
   ,, amplitude ... ... 75° N. of E.
3. LOCKYER, 1891—
   Magnetic azimuth of axis ... ... N. 23° E.
   ,, amplitude ,, ... 67 N. of E.
   Assumed correction $4\frac{1}{2}°$
   Astronomical amplitude ... ... $71\frac{1}{2}°$

As my value agrees closely with that of Biot, I adhere to it; and it gives, for the amplitude of the temple of Isis at right angles to the Hathor temple, 18½° S. of E.

Now, it is stated distinctly in the inscriptions that "the place of the birth of Isis is to the north-west of the temple of Hathor, its portal is turned to the east, and the sun shines on its portal when it rises to illuminate the world."[1] We learn from this that the small temple was locally celebrated as the birthplace of Isis.

It is, then, a temple of Isis. Who was Isis?

Let us begin by considering the temple, remarking that the inscriptions, apparently relating to both temples, are found in one only. On this point, I, for the present, content myself with quoting Plutarch's statement[2] that Isis and Hathor were the same divinities—at all events, in later Egyptian times.

If we study the inscriptions—and this, thanks chiefly to Mariette's magnificent book on Denderah, we can do—we find that they give out a very certain sound. Here is one of them:—

> "She [*i.e.* her Majesty Isis] shines into her temple on New Year's Day, and she mingles her light with that of her father Rā on the horizon."

Here we have nothing more nor less than a distinct and perfectly accurate statement relating to the cosmical rising of a star, *i.e.*, as I have before explained, of the sun and the star both rising at the same instant of time.

Further, in the inscriptions the "*rising of Hathor*" is mentioned distinctly. "La grande déesse Sefekh [Sesheta] apporte les écrits qui se rapportent à ton lever, ô Hathor, et au lever de Rā."[3] Everybody knows that "Rā" means the sun, and

[1] Mariette, "Denderah," vol. i., p. 263.

[2] Mariette, *op. cit.*, p. 142. Plutarch wrote in the first century A.D.

[3] Mariette, *op. cit.*, p. 206.

RUINS OF THE MAMISI (PLACE OF BIRTH) OR TEMPLE OF ISIS AT DENDERAH.

therefore the rising of Rā is at once accepted by everybody as obviously meaning sunrise. But if we find "Hathor" treated in the same way as the sun, then Hathor must be a celestial body rising like the sun. I consider this a very important conclusion to arrive at, for many reasons.

But, further, Hathor was also worshipped, according to the inscriptions,[1] under the name of Sothis.

Now we know, quite independently of all mythology, that Sothis is simply the Greek form of the Egyptian name (Sept) of the star Sirius.

Taking, then, all these inscriptions together, we have an absolute astronomical demonstration of the fact that the "rising of Hathor," which is referred to mythologically in the inscriptions given by Mariette, was the rising of Sirius; that the star which "shone into the temple, and which mingled her light with the light of her father Rā," was really the star Sirius. We get the demonstration of the fact that mythologically the star Sirius was Hathor, or otherwise Isis.

In other words, we find a star personified; Sirius being personified as Hathor or Isis.

But we can go much further than this. It is possible, as I have shown, to determine the position of Sirius in past times, and therefore to determine whether the light of that star ever did fall along the axis of the temple. We know its orientation approximately—18½° S. of E.—so that any celestial body which rose at that amplitude would shine upon any object enshrined in the sanctuary. In the case of Sirius, the conditions are such that, owing to the precessional movement, the distance of the star from the equator has been gradually lessening from the earliest times. Its declination in 8000 B.C. was 50° S.; it became something more than 17° S. in A.D. 1000.

[1] Mariette, *op. cit.*, p. 142.

Knowing the declination, it is easy to determine the amplitude—and given the conditions at the temple of Isis at Denderah, viz., that we are practically dealing with a sea horizon, we find that the temple really pointed to Sirius about 700 B.C., which is the date Biot found for the construction of the zodiac in the temple of Osiris, referred to in Chapter XIII.

Further, it is easy to show that Sirius at that date rose with the sun on the Egyptian New Year's Day;[1] in mythological language, she mingled her light with that of her father Rā on the great day of the year.

As this is the first instance of such personification that we have come across, it behoves us to study it very carefully. Why was Sirius personified and worshipped?

The summer solstice—that is, the 20th of June, the longest day—was the most important time of the Egyptian year, as it marked the rise of the all-fertilising Nile. It was really New Year's Day. It has been pointed out, times without number, that the inscriptions indicate that by far the most important astronomical event in Egyptian history was the rising of the star Sirius at this precise time.

Now it seems as if among all ancient peoples each sunrise, each return of the sun—or of the sun-god—was hailed, and most naturally, as a resurrection from the sleep—the death—of night: with the returning sun, man found himself again in full possession of his powers of living, of doing, of enjoying. The sun-god had conquered death; man was again alive. Light and warmth returned with the dawn in those favoured Eastern climes where man then was, and the dawn itself was a sight, a sensation, in which everything conspired to suggest awe and gratitude, and to thrill the emotions of even uncivilised man.

[1] Hathor is termed "La maîtresse du commencement de l'an." Mariette, *loc. cit.*, p. 207.

What wonder, then, that sunrise was the chief time of prayer and thankfulness? But prayer to the sun-god meant, then, sacrifice; and here a practical detail comes in, apparently a note of discord, but really the true germ of our present knowledge of the starry heavens which surround us.

To make the sacrifice at the instant of sunrise, preparations had to be made, beasts had to be slaughtered, and a ritual had to be followed; this required time, and a certain definite quantity of it. To measure this, the only means available then was to watch the rising of a star, the first glimmer of which past experience had shown to precede sunrise by just that amount of time which the ritual demanded for the various functions connected with the sunrise sacrifice.

This, perhaps, went on every morning, but beyond all question the most solemn ceremonial of this nature in the whole year was that which took place on New Year's morning, or the great festival of the Nile-rising and summer solstice, the 1st of Thoth. Besides the morning ceremonial there were processions of the gods during the day.

How long these morning and special yearly ceremonials went on before the dawn of history we, of course, have no knowledge. Nor are the stars thus used certainly known to us. Of course any star would do which rose at the appropriate time before the sun itself, whether the star was located in the northern or in the southern heavens. But in historic times there is no doubt whatever about the star so used. The warning-star watched by the Egyptians at Thebes, certainly 3000 B.C., was Sirius, the brightest of them all, and there is complete evidence that Sirius was not the star first so used.[1]

[1] "Besides the solstice and the beginning of the Nile flood, there was an event in the sky which was too striking not to excite the general attention of the Egyptian priesthood. We also know from the newly-discovered inscriptions from the ancient empire that the risings of Orion and Sirius were already attentively followed and mythologically utilised at the time of the building of the pyramids."—KRALL.

CEREMONIAL PROCESSION IN AN EGYPTIAN TEMPLE. *(From a Restoration by the French Commission.)*

The astronomical conditions of the rising of this star have, fortunately for us, been most minutely studied both by Biot and, in more recent times, by Oppolzer, and from their labours it seems to be abundantly clear that the rising of Sirius at the solstice was carefully watched certainly as early as 3285 B.C., according to Biot's calculations; and, further, that the rising of the same star was still studied in a relatively modern time. At the earlier date its heliacal rising was observed, but in later times means had been secured of noticing its cosmical rising, because although it rose long before the sun on the longest day 3000 B.C., it rose *with* the sun on the same day in the later times referred to. This "cosmical rising" observation was doubtless secured by the construction of their temples, as I have shown.

We are, then, astronomically on very firm ground indeed. We have got one step into the domain of mythology. I assume it is agreed that we have arrived at the certain conclusion that the goddess Hathor or Isis personified a star, Sirius, rising at the dawn; and that the temple of Isis at Denderah was built to watch it.

CHAPTER XX.

THE PERSONIFICATION OF STARS (CONTINUED)—THE TEMPLE OF HATHOR AT DENDERAH.

In Chapter XVII. I quoted from the inscriptions relating to the alignment of the axis of the temple of Hathor at Denderah. It will be remembered that the king, while stretching the cord, had his glance directed to the *āk* of the constellation of the Thigh. Further, we saw in the last chapter that the amplitude of the temple axis is $71\frac{1}{2}°$ N. of E.

A copy of Biot's plan giving his value of the orientation is given on the next page.

I have shown how truly the temple of Isis was pointed to Sirius. We have now to try to find a star towards which the temple of Hathor may have been pointed in like manner.

It will be generally understood that in an inquiry of this kind there are very many difficulties, chiefly depending upon the uncertainty of the building-date of the original foundation, and upon the indeterminate nature of the information available. But although we meet with these difficulties in the case of the temple of Hathor, there are many from which we are free. In the case of many of the temples in Egypt we have no knowledge of the tutelary divinity. For a great many temples no observational data exist; they have not been properly measured—that is, we do not know exactly in what direction they point or what their amplitudes are; and, further we do not know anything of the horizon at the temple building, so as to be able to make the necessary corrections due to heights of hills.

This premised, I will now return to the statement regarding the temple of Hathor, to see what can be made of it on the view that either the middle or the chief point,

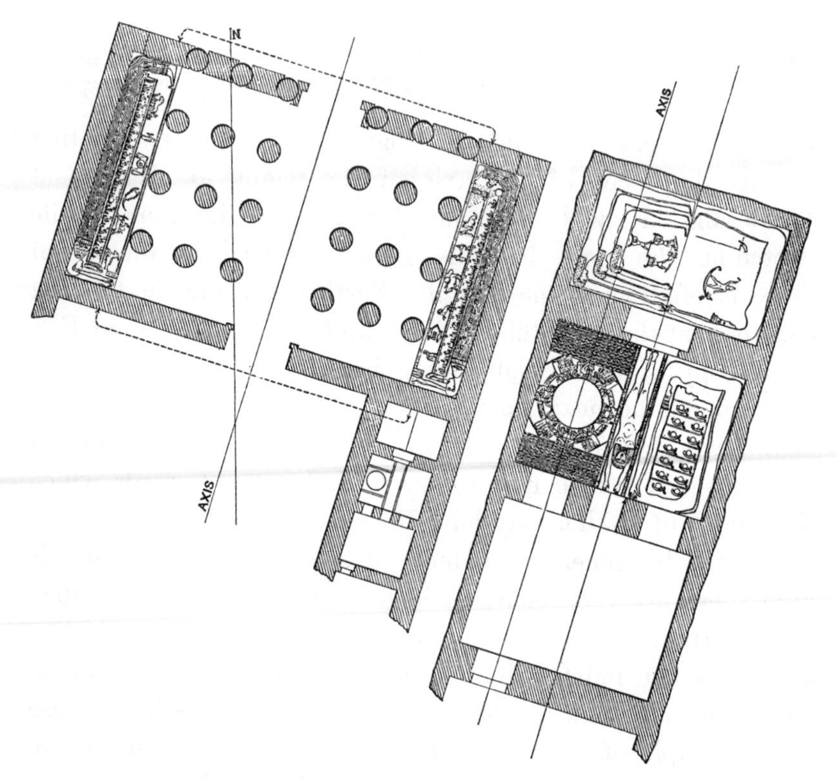

ORIENTATION OF THE TEMPLE OF HATHOR AT DENDERAH (BIOT). (THE TEMPLE OF OSIRIS ON THE ROOF IS ALSO SHOWN ON A LARGER SCALE.)

that is, the brightest star, of the constellation of the Great Bear as we now know it, was the one referred to, and that the cord was stretched to the star on the horizon.

The first question which arises is, Was there any reason why δ Ursæ Majoris at the centre, or *a* the brightest, should have been used as the orientation point at any time? Was there any reason why any special sanctity should have been associated with either? Certainly not, in the case of δ, on account of its magnitude, because Dubhe, not far from it, is much brighter; and possibly not, in the case both of δ and α, on account of the time of their heliacal rising. We seem therefore in an *impasse* along this line of inquiry; but a further consideration of the question brings out the remarkable fact that at three widely-sundered points of time the stars α Lyræ, α Ursæ Majoris, and γ Draconis have been the brightest stars nearest the North Pole, and with such declinations that α Lyræ would be visible at one of the dates, α Ursæ Majoris at another, and γ Draconis at another still—all rising in nearly the same amplitude far to the north.

In Chapter XVIII. I have shown that one of the temples, and possibly a series of them, at Thebes were directed to γ Draconis. It is interesting, then, to carry the inquiry further. It may possibly explain how it is that we get a definite statement about the *āk* of the Great Bear in one case and a certain sure orientation to γ Draconis in the other.

In the first place, it has to be borne in mind that when a star is circumpolar—that is, never sets—no temple can be directed to its rising. Now, accepting the *āk* as the brightest star (and as I stated in Chapter XVII., it might, indeed, have been the central one as well in the old constellation, for we do not know its limits), we have to deal with the facts concerning α Ursæ Majoris, called by the Arabians Dubhe.

The latitude of Denderah is a little over 26° N., therefore all stars with a less polar distance than that—or, to put it another way, all stars with a declination greater than (90°−26°=) 64° N.—will be circumpolar. Now, the declination of Dubhe was greater than 64° between 4000 B.C. and 1500 A.D. (I neglect refractions and hills); hence, if there is any truth in the statements made in the building ceremonials, the temple could not have been founded between those dates.

But what are the records concerning this temple? We know that the structure as we see it was built in the time of the last Ptolemies and the first Roman emperors, and I have already shown that at those dates the Great Bear (the old Thigh) did not rise at all, as it was circumpolar.

It is also known that there was a temple here in the time of Thothmes III., and even earlier, going back to the earliest times of Egyptian history. King Pepi, of the Sixth Dynasty (*circ.* 3233 B.C.), is portrayed over and over again in the crypts.

Even this is not all the evidence in favour of a high antiquity. In one of the crypts (No. 9), according to Ebers and Dümichen, there are two references to the earliest plans of the temple. One inscription states that the great ground-plan (*Senti*) of Ant (Denderah) was found in old writing on parchments of the time of the followers of Horus (sun-worshippers) preserved in the walls of the temple during the reign of King Pepi. Another inscription goes further, referring to the restoration by Thothmes III. (*circ.* 1600 B.C.) of the temple to the state in which it was found described in old writings of the time of the King Chufu (Cheops) of the Fourth Dynasty (*circ.* 3733 B.C.). If any faith is to be placed in this inscription, it seems to me to suggest a still higher antiquity. There would have been more reason for describing an antique shrine than a brand-new one.

Still another inscription runs:—

"King Tehuti-mes III. has caused this building to be erected in memory of his mother, the goddess Hathor, the Lady of An (Denderah), the eye of the Sun, the heavenly queen of the gods. The ground-plan was found in the city of An, in archaic drawing on a leather roll of the time of the Hor-Shesu: it was found in the interior of a brick wall in the south side of the temple in the reign of King Pepi." [1]

But let us see what the facts are regarding the date supplied by the temple itself, accepting the statement made regarding the actual operations at the laying of the foundation stone originally.

To determine the dates approximately, we find that an amplitude of $71\frac{1}{2}°$ N. of E. in the latitude of Denderah gives a declination of $57\frac{3}{4}°$ N., with a sea horizon (correcting for refraction) $58\frac{3}{4}°$ N. with hills 1° high, and $59\frac{3}{4}°$ N. with hills 2° high, which is not far from the exact conditions.

The star Dubhe had the declination of 60° N. in 5000 B.C.

If, then, I am right in my suggestion as to the word *āk* referring to *a* Ursæ Majoris, we find the closest agreement between the astronomical orientation; the definite statement as to a certain star being used in the building ceremonies; the inscriptions in the crypts referring to Cheops as the earliest historical personage who describes the building, and to the Shesu-Hor as the original designers of the building. According to most authorities, 5000 B.C. lands us in the times of the Shesu-Hor before Mena.

I must confess that this justification of the double record strikes me as very remarkable, and I think it will be generally conceded that further local observations should be made in order to attempt to carry the matter a stage beyond a first approximation.

We have got so far, then. If we take the history as we find it, and further take the trouble to work out the very definite statements made, we find that the temple was founded pointing to the rising of Dubhe before it became circumpolar, and that in those times this star was symbolised by the name of Hathor.

[1] Brugsch, "Egypt," Edition 1891, p. 189.

We may accept, then, the possibility that as the temple of Isis was oriented to Sirius, that dedicated to Hathor was directed to Dubhe.

It will have been obvious from what has preceded, that if the worship of Hathor was to go on at all, and if it were in any way connected with the observations of a star rising near the north point of the horizon, a new star must be chosen when *α* Ursæ Majoris became circumpolar. That is the first point.

I have already stated that *α* Ursæ Majoris began to be circumpolar at Denderah 4000 B.C. I may now add that γ Draconis ceased to be circumpolar about 5000 B.C. They had the same declination (62° N.) and the same amplitude (78° N.) 4400 B.C.

Mariette's plan shows a second temple oriented to N. 6° E., which we may perhaps be justified in taking as N. 9° E., since his azimuth of the great temple differs from Biot's and my own by 3°.

The corresponding declination would be 63° N. of E., the declination of Dubhe in 4200 B.C. and of γ Draconis in 4300 B.C. The temple may well, therefore, have been erected when both stars had the same amplitude, the apparent difference of 100 years being due to the uncertainty of the measures available.

The second point, then, is that when Dubhe, which, while it rose and set, was the brightest star near the pole which did so, *became circumpolar;* γ Draconis, when it *ceased to be circumpolar*, fulfilled these conditions; astronomically, then, it became the natural successor of *α* Ursæ Majoris.

I have before pointed out that it is not impossible that a temple once oriented to a certain star, and long out of use on account of the precessional movement, may be utilised for another, and be rehabilitated in consequence, when that same movement brings another conspicuous star into the proper rising amplitude.

This consideration at once leads to my third point, which is, that after Dubhe became circumpolar the temple of Hathor at Denderah would become useless—there would be no star to watch—unless a new star was chosen.

Now, let us suppose this to have been so, and that the natural successor of the star in question were chosen. Studying the facts as before approximately, as final data are not yet available, we have the declination $59\frac{3}{4}^{\circ}$ N. This was the declination of γ Draconis about 3500 B.C., assuming hills 2° high, which I think is too much; 3300 B.C., with hills $1\frac{1}{2}^{\circ}$ high.

In the present case the orientation fits γ Draconis in the historic period, but it also fits Dubhe in the times of the Hor-shesu, the dimly-seen followers of Horus, or sun-worshippers, before the dawn of the historic period.

Next let us go back to the inscriptions. We found that King Pepi is portrayed over and over again in the crypts, and, which is more important, that the plan of the temple on parchment, dating from the times of the Shesu-Hor, had actually been walled up in the temple during the reign of the same king, no doubt at the ceremony of restoration or laying a new foundation stone, as is sometimes done to this day.

Now, Pepi's date, according to the chronologists, is 3200 B.C., a difference of 100 years only from the rough orientation date.

We see, therefore, the full importance of the work done in Pepi's reign. The *āk* of the Thigh was no longer of use; but a new star was now available. Hathor was rehabilitated. Perhaps even the priests alone knew that the star had been changed.

By the temple of Hathor, then, if we assume that the

record is absolutely true (and I, for one, believe in these old records more and more), and that Cheops only described a shrine founded by the Hor-shesu, we are carried back to *circ.* 5000 B.C. I am indebted to my friend Dr. Wallis Budge for the suggestion that the position of Denderah as the terminus of the highway from the Red Sea—which may soon again be reached by a railway from Keneh to Kosseir!—would have made it one of the most important places in ancient Egypt.

It is important to note that at a very early date the traffic between the Nile Valley and the Red Sea, and thence probably with Arabia and South Africa, flourished, and grew to be a by no means insignificant commerce.

According to Ebers,[1] "the oldest and most famous of all these highways is that which led from Koptos (Keneh, Denderah) to the Red Sea, through the valley now known as the Wady Hammamāt, and called by the ancient Egyptians Rohanu. It was a busy high-road, not alone for trading caravans, but from time to time for stonemasons and soldiers, whose task it was to hew the costly building materials from the hard rocks, which here abound, and to prepare the vast monoliths which were finished *in situ*, and then to convey them all to the residence of the Pharaohs. A remarkably beautiful kind of alabaster, of a fine honey yellow or white as snow, is found in these mountains." Another road led from Esneh or Edfû to the ancient port of Berenice. We shall see in the sequel that the temple of Redisieh on this route was dedicated to the same cult as that at Denderah.

If the above results be confirmed, we have a most definite indication of the fact that in the rebuilding in the times of Pepi, Thothmes III., and the Ptolemies, the original orientation

[1] Ebers, "Egypt," p. 335.

of the building was not disturbed; and that in the account of the building ceremonies we are dealing as surely with the laying of the first foundation-stone as with the original plan.

In any case the consideration has to be borne in mind that the series of temples with high northern (and southern) amplitudes at Denderah, Thebes, and possibly other places, were nearly certainly founded before the time at which the heliacal rising of Sirius, near the time of the summer solstice, was the chief event of the year, watched by priests, astronomers—if the astronomers were not the only priests—and agriculturists alike. Now we know, from Biot's calculations, that this became possible *circ.* 3285 B.C., and that Sirius—though, as I am informed by Prof. Maspero, *not* its heliacal rising—is referred to in inscriptions in pyramid times.

Subsequent research may possibly show that these temples had to do with the heralding of sunrise throughout the year, the Sirian temples being limited to New Year's Day.

CHAPTER XXI.

STAR-CULTS.

THE last two chapters, then, have brought us so far There are two principal temples at Denderah. The smaller is called the temple of Isis. It is oriented 18½° S. of E. The inscriptions tell us that the light of Sirius shone into it, and that Sirius was personified as Isis. We can determine astronomically that the statement is true for the time about 700 B.C., which was the date determined independently by Biot for the circular zodiac referred to on page 18.

The larger temple is called the temple of Hathor. It is oriented 71½° N. of E. The inscriptions very definitely tell us what star cast its light along its axis, and give also definite statements about the date of its foundation, which enable us to determine astronomically that in all probability the temple was oriented to Dubhe somewhat later than 5000 B.C.

Now we are *certain* that Isis personified Sirius. That "Her Majesty of Denderah" was Sirius, at all events in the later times referred to in the inscriptions, is not only to be gathered from the inscriptions, but has been determined astronomically.

It is also *probable* that Hathor personified Dubhe. Now this looks very satisfactory, and it seems only necessary to test the theory by finding temples of Isis and Hathor in other places, and seeing whether or not they were oriented to Sirius and Dubhe respectively.

But, unfortunately for us, we have already learned from Plutarch that Isis and Hathor are the same goddesses, although they certainly personify different stars, if they personify stars at all.

We seem, then, in a difficulty, and at first sight matters do not appear to be made any clearer by the fact that Hathor (and, therefore, Isis) was worshipped under different names in every nome.

Lanzoni, in his admirable volumes on Egyptian mythology, gives us, not dealing with the matter from this point of view at all, *no less than twenty-four variants for Hathor!*

In the temple at Edfû no less than 300 names are given with the various local relations and forms used in the most celebrated shrines.[1]

In the inscriptions at Denderah itself a great number of variants is given.[2] It is important to give some of them in this place; the full value of the information thus afforded will be seen afterwards.

Hathor	of Denderah	=		
Hathor	of Denderah =	Sekhet	of	Memphis.
,,	,,	Neith	,,	Sais.
,,	,,	Saosis	,,	Heliopolis.
,,	,,	Nehem-an	,,	Hermopolis.
,,	,,	Bast	,,	Bubastis.
,,	,,	Bes-t		
,,	,,	Anub-et	,,	Lycopolis.
,,	,,	Amen-t	,,	Thebes.
,,	,,	Bouto	,,	Unas.
,,	,,	Sothis	,,	Elephantine.
,,	,,	Apet		
,,	,,	Mena-t		
,,	,,	Horus (female)	,,	Edfû.

One variant is of especial importance in the present connection, and is emphasised in a special inscription in one of the chambers of the temple of Hathor—*not*, be it remarked, in the temple of Isis.

"Elle est la Sothis de Denderah, qui remplit le ciel et la terre de ses bienfaits. Elle est la régente et la reine des villes. . . . Au Sud elle est la reine du

[1] Mariette, pp. 168 and 178.

[2] Dümichen, "Bauurkunde der Tempelanlagen von Dendera," p. 20.

maître divin; au nord elle est la reine des divins ancêtres. Rien n'est établi sans elle. . . . Elle est la grande dans le ciel, la reine parmi les étoiles."[1]

Well may Mariette remark on this:

"Cette invocation à Sothis, dans une chambre consacrée à la consécration de certains produits de la terre, n'a rien qui doit surprendre. Sothis est le symbole du renouvellement de l'année et de la résurrection de la nature. Au lever héliaque de Sothis, le Nil sort de son lit. Jusqu'à ce moment la terre de l'Égypte est stérile et nue. Fécondée par la fleuve, elle va se couvrir d'une verdure nouvelle."

But the Sothis here in question is *Sirius*, the star to the rising of which the temple of Isis, and *not* the temple of Hathor, was directed!

We have, then, at Denderah a temple *not* pointed to Sirius, the worship in which is that of Hathor, and there can be little doubt that we have astronomically determined the fact that "Her Majesty of Denderah" was really the star Sirius.

We can pass from Denderah to the temple of Hathor at Thebes. The general plan of Thebes prepared by Lepsius indicates the orientation of the temple of Dêr el-Bahari, to which I refer, the temple in the western hills of Thebes, embellished by Queen Hatshepset (*circ.* 1600 B.C.). This temple, instead of being oriented $71\frac{1}{2}°$ N. of E., lies $24\frac{1}{2}°$ S. of E.; it can never, therefore, have faced the star observed in the temple of Hathor at Denderah. There is also another temple annexed to the temple of Amen-Rā, which received the light of Sirius in former years. These temples were, in all probability, intended to observe the same star which was subsequently observed in the temple of Isis at Denderah.

That is one point; here is another. We have it from Plutarch[2] that Isis=Mut=Hathor=Methuer.

The amplitude of the temple at Denderah dedicated to

[1] Mariette, p. 156.

[2] "Isis and Osiris," Parthey, cap. 56.

Hathor is $71\frac{1}{2}°$ N. of E. (59° N. declination). That of the temple dedicated to Mut at Karnak is $72\frac{1}{2}°$ N. of E. ($58\frac{3}{4}°$ N. declination), which, assuming for a moment the same star to have been used, corresponds to a date (according to the height of the horizon) of *circ.* 3000 to 3500 B.C. This is therefore later than the original foundation of the Hathor temple of Denderah, but not far from the date of its restoration by Pepi.

It is fundamental to the orientation theory that the cult shall follow the star. But we have here the same cult, according to Plutarch; we are hence permitted to suggest that in dealing with the temples of Hathor at Denderah and Mut at Thebes we are dealing with local names of the same goddess personifying the same star.

Two lines of argument may be followed to strengthen this conclusion.

The first has to do with the orientation of the temple of Mut at Thebes. There is no statement of its great antiquity, as in the case of the temple of Hathor at Denderah. Here we find again one of the great difficulties in our way, the impossibility of running back to the original foundation among the many restorations effected of the most important among the Egyptian temples. The temple of Mut is ascribed to Amen-hetep III., but I cannot hold this to be the original foundation, for the following reasons:—

1. With its orientation in the time of Amen-hetep III. it pointed to no star in particular.

2. There is a series of four temples at Thebes turned to the same part of the horizon nearly, their amplitudes ranging from 62° to $72\frac{1}{2}°$ N. of E. Of these temples that of Mut has the highest amplitude; the one with the lowest but one is the temple lettered M by Lepsius. There is no question

about the real founder of this temple, and there is not much question as to the date of the founder, Rameses III.

Now in the time of this king a temple erected with the orientation given pointed precisely to γ Draconis. (*See* Chapter XVIII.) The amplitude was 62° N. of E.; the time, 1200 B.C. If we take the simplest case in the orientation theory—that the amplitudes

62° N. of E.
$63\frac{1}{2}$° "
$68\frac{1}{2}$° "
$72\frac{1}{2}$° "

were given to the various temples to enable observations to be made of the same star, which was being carried nearer the equator by the precessional movement, we can not only date the temple of Mut, but find an explanation of Plutarch's equation Hathor = Mut.

In other words, we watch the Mut-Hathor worship provided for from 3000 B.C. to the times of the Ptolemies.

So that here we have a very concrete case of the cult following the star, not only in the same place, but at different places, and we are driven to the conclusion that Hathor at Denderah and Mut at Thebes, exoterically different goddesses, were esoterically the same star, γ Draconis.

We are not, however, limited to a comparison between Denderah and Thebes. We have Annu and Abydos, and other places, to appeal to, since there are temples remaining there also facing N.E. Those at Abydos, however, we must leave out of consideration here, as their exact orientation is not determined. With regard to Heliopolis, and dealing with the obelisk which tradition tells us was erected by Usertsen I., the orientation of its N.E. face, according to my own observations, taking the present variation at $4\frac{1}{2}$° W., is 77° N. of E.

This corresponds approximately to a declination of $57\frac{1}{2}°$ N., which was the declination of γ Draconis in 2500 B.C. The date given to Usertsen I. by Brugsch is 2433 B.C.

This is very satisfactory so far, but we can go further. Here we are landed evidently in the worship of one of the local divine dynasties, that of Set; and we may justly, therefore, ask if Usertsen did not do at Heliopolis what it is very probable Pepi did at Denderah—namely, embellish an old temple which had in the first instance been used for observations of Dubhe and appropriate it to the use of the new Hathor γ Draconis. If this were so, then the original foundation stone was laid about 5100 B.C.

The next line of argument is furnished by the emblems which are associated with the various goddesses. These obviously indicate that they arose in a time of totemism, when each tribe or nome had its special totem, which would be certain to be associated with the local goddesses or the stars which they personified.

The local totem of the special warning-star in use at any time or place may be anything: hippopotamus, crocodile, hawk, vulture, lion, or even some other common living thing into which the totem degraded when the supply of the original fell short.[1]

Hence, as the number of warning-stars was certainly very restricted, they—or, rather, the goddesses which typified them—had different names in almost every nome. Hence Egyptian mythology should be, as it is in fact, full of synonyms; each local name being liable to be brought into prominence at some time or another, owing to adventitious circumstances relating either to dynasties or the popularity of some particular shrine.

[1] Have we such instances of degradation in the cat replacing the lion and the black pig the hippopotamus, to give two instances?

Applying this test of symbolism, we find in the case of Hathor that the symbolism was double.

The Denderah *Hathor* was connected with the hippopotamus, while at Thebes *Mut* was represented by a hippopotamus.

CAPITAL, WITH MASKS OF HATHOR WITH COWS' EARS.

Now this symbol of the hippopotamus helps us greatly, because it allows evidence to be gathered from a consideration of the old constellations. I do not think it is saying too much to remark that among these the attention of the North Egyptians was almost exclusively confined to the circumpolar ones. Further, the mean latitude being, say, 25°, the circumpolar region was a restricted one; 50° in diameter, instead of over 100°, as with us. But not quite exclusively, for to them in later times, as to us now, the Great Bear and *Orion* were the two most prominent constellations in the heavens; for them, as for us, they typified the northern and southern regions of the sky.

There can be no question that the chief ancient constellation in the north was the Great Bear, or, as it was then pictured, the Thigh (Mesχet). After this came the Hippopotamus. I had come to the conclusion that this had been replaced on our maps by part of Draco before I found that Brugsch and Parthey had expressed the same opinion.

The female hippopotamus typified Taurt, the wife of Set (represented by a jackal with erected tail, or hippopotamus), and one of the most ordinary forms of Hathor is a hippopotamus. There is evidence that the star we are considering, γ Draconis, occupied the place of the head or the mythical headgear.

Here, then, in the actual symbolism of Hathor we find γ Draconis as distinctly pointed to as by the orientation of the temples.

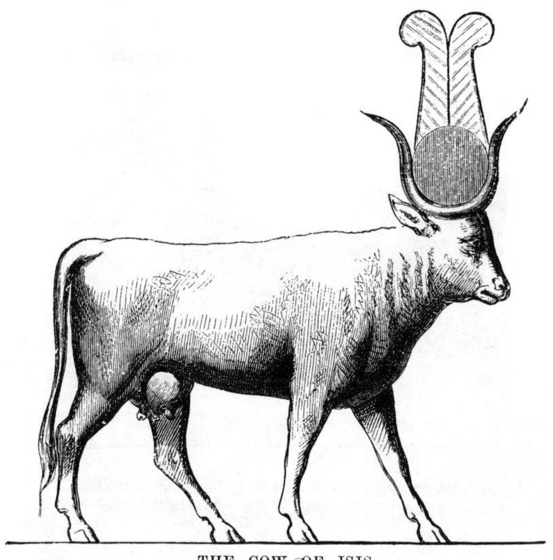

THE COW OF ISIS.

The other symbolism is quite different; instead of a hippopotamus we deal with a cow.

In the inscriptions at Denderah we find the star Sirius represented by a cow in a boat. In the circular zodiac we have the cow in the boat, the point of the beginning of the year, and the constellation Orion, so located as to indicate clearly that, at that time, the beginning of the year fell between the heliacal rising of Sirius and of the stars in Orion. Sirius was Isis-Sothis.

If we go to Thebes, we pass there from the cow Isis-Sothis to Isis-Hathor, and there we find the mythology retains the idea of the cow, the cow gradually appearing from behind the western hills. There is not a doubt, I think, that the basis of this mythological representation was, that the temple which was built to observe the rising of the star at a time perhaps

HATHOR AS A COW.

somewhat later than that given by Biot (3285 B.C.) was situated in the western hills of Thebes, so that Hathor, the goddess on which the light was to fall in the sanctuary, was imaged as dwelling in the western hills. At Philæ we get no longer either Isis-Sothis or Isis-Hathor, but Isis-Sati.[1]

[1] It has been assumed by several authorities in Egyptology that Sati is a variant for Sirius. It is quite certain that in late times there was a temple at Philæ oriented to Sirius; but there are many grounds for supposing that both Sati and Anuqa referred to special southern stars. There were several points of dissimilarity between Philæ (and Elephantine) and Thebes.

Now just as certainly as the hippopotamus had to do with the constellation Draco, the cow had to do with Sirius, for Sirius was represented as a cow in a boat.

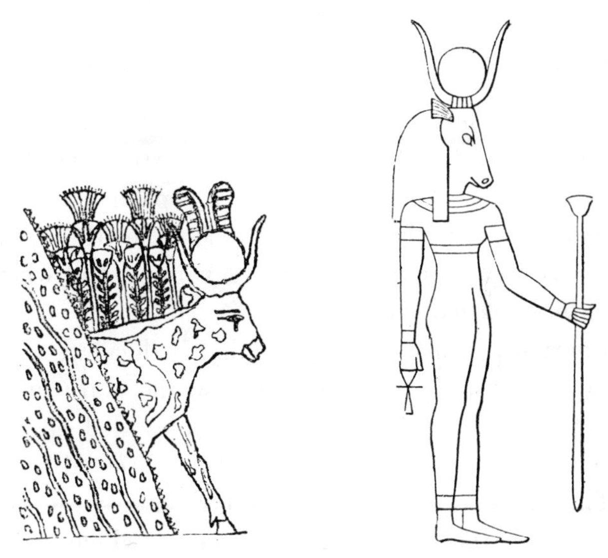

HATHOR, "THE COW OF THE WESTERN HILLS."

It may be gathered from this how truly astronomical in basis was the mythologic symbolism to which we have been driven in the effort to obtain more light; and, indeed, it is necessary for us to consider it still more closely.

CHAPTER XXII.

STAR-CULTS (CONTINUED)—AMEN-T AND KHONS.

WHEN I had the privilege of discussing at Thebes the orientation hypothesis with M. Bouriant, the distinguished head of the French School of Archæology in Egypt, he suggested that I should accompany him one day to Medînet-Habû, at which place he was then superintending excavations, and where there are three temples dedicated to Amen.

M. Bouriant, from the first, saw that if there were anything in the new views, the cult must follow the star; and it was natural, therefore, that the three temples dedicated to the same divinity at the same place should be directed to the same star. The three temples to which I refer are the two well-known temples the lack of parallelism of which has been so often remarked, and a third much smaller one, built more recently, lying to the south-west. The amplitudes I found to be as follows:—

	Amplitude S. of E.
Ethiopian or Ptolemaic Temple	45°
Great Temple	46½°
Ancient Temple	51½°

On the orientation hypothesis we were dealing with a star the S.E. amplitude of which was decreasing like that of Sirius; *it was therefore in the same quarter of the heavens.*

But which star? To investigate this it was best to deal in the first instance with the orientation of the great temple, since its building date was supposed to be that most accurately known; and there is not much danger in doing this in the present case, because the king obviously had not expanded an old temple, for there it still is alongside.

The king was Rameses III., the date, according to Brugsch, 1200 B.C., and the hills to which the temples are directed may be taken as 1° high. With these data we get the declination appropriate to the amplitude of the temple 40° S. Now, this was nearly the declination of the star Phact or *α* Columbæ in the time of Rameses III.; the orientation date is 1250 B.C.

Taking this star, then, and correcting for heights of hills and refraction, we get approximately the following dates:—

	B.C.
Modern Temple	900
Great Temple	1250
Ancient Temple	2525

If the hills are taken as 1½° high, these dates will stand 750, 1150, and 2400.

The date 700 B.C. we have already found as the probable date of the undertaking of the restoration at Denderah. It is the time of the victorious march of the Theban priests northwards from their exile at Gebel Barkal.

The date 2400 B.C. lands us in the times of the great solstitial king, Usertsen I., about whom more in a subsequent chapter. Although the more ancient temple is generally ascribed to Thothmes III., traces of the work of Amen-hetep I. have been discovered. I think we have a case here where the eighteenth dynasty enlarged and embellished a shrine erected by the twelfth dynasty, precisely as the temple of Amen-Rā at Karnak has been traced back to the twelfth dynasty.

If I am right, then, it follows that temples erected to stars associated in any way with the chief cult, such as that of Amen-Rā, may either be dedicated to the god or goddess personified by the star or to the associated solar deity. Thus at

Thebes we have the temple of Mut, so-called, though Mut was the wife of Amen-Rā; and the temples now under consideration, called temples of Amen, though they are dedicated to the goddess Amen-t, the wife of Amen. This may or may not be connected with the fact that the first of them was dedicated possibly before the cult of Amen alone had been intensified and expanded by the Theban priests—probably in the eighteenth dynasty—into the cult of the solstitial sun-god Amen-Rā.

There is evidence, indeed, that Amen-t replaced Mut in the Theban triad. With regard to these triads, a few words may be said here from the astronomical point of view, though the subject, I am told, is one on which a great diversity of opinion exists on the part of Egyptologists.

I have collected all the most definite statements I can find on this head, and it is certainly interesting to see that in many cases, though not in all, the triad seems to consist of a form of the sun-god, together with two stellar divinities, one of them certainly associated with the heliacal rising of the sun at some time of the year, and therefore a recognised form of Isis or Hathor. Thus we have:—

Place.	Triad.
Thebes	Amen-Rā
(Greater Triad)	Mut
	χonsu
(Lesser Triad)	χem-Rā
	Tamen (? Amen-t)
	Harka
Denderah	Atmu
	Isis
	Hathor
Memphis	Atmu
	Sekhet
	Ptah
Hermonthis	Menθu-Rā
	Ra-Ta (= Hathor)
	Hor-Para

Not only may this table enable us to see how Amen-t was sunk at Medînet-Habû in the term Amen, but it enables us to consider a similar case presented by those temples at Thebes, some of them associated with Khons and another with Amen, referred to in Chapter XVII.

The temple of *Khons* is among the best known at Karnak; the visitor passes it before the great temple of Amen-Rā is reached. M. Bouriant was able to prove, while we were together at Karnak, that the temple of Seti II., nearly parallel to it, was also dedicated to *Khons;* but the temple B of Lepsius, nearly parallel to both, is sacred to *Amen.* It is seen at once that the main cult is the same, although the amount of detail shown in the reference is different—we have the generic name of the triad in one case, the specific name of the member of the triad in the other.

As this is the first time a setting star has been in question, it is well to point out that in this case the ancient Egyptians no longer typified the star as a goddess but as a god—and, more than this, as a dying god; for Khons is always represented as a *mummy*—the Osiris form. Egyptologists state that both Thoth and Khons were moon-gods. Perhaps the lunar attributes were assigned prior to the establishment of sun-worship.

I shall show, subsequently, that the temples now being considered find their place in continuous series stretching back in the case of Amen-t to 3750 B.C., and in the case of Khons to possibly a long anterior date.

In the case of Amen-t and Khons, therefore, where we are free from the difficulties connected with the interchange of the titles of Isis and Hathor at Denderah, the star-cults stand out much more clearly, and we get a step further into the domain of mythology.

But what did the cults mean? What was the utility of

them? What their probable origin? The cult of Sirius we already understand.

I will deal with Amen-t first. No doubt it will have been already asked how it came that such an unfamiliar star as Phact had been selected.

Here the answer is overwhelming. This star, although so little familiar to us northerners, is one of the most conspicuous of the stars in the southern portion of the heavens, *and its heliacal rising heralded the solstice and the rise of the Nile before the heliacal rising of Sirius was useful for that purpose!*

In Phact we have the star symbolised by the ancient Egyptians under the name of the goddess Amen-t or Teχi, whose figure in the month table at the Ramesseum leads the procession of the months.

Amen-t, the wife of the solstitial sun-god Rā, symbolised the star the rising of which heralded the solstice; and the complex title Amen-Rā signified in ancient times, to *those who knew*, that the solstitial sun-god Rā, so heralded, was meant.

The answer is clear, though not so simple in the case of Khons. The setting of Canopus marked the autumnal equinox about 5000 B.C. We have found that the first Khons temple at Karnak was possibly built as late as 2000 B.C., when the utility of the observations of Canopus from this point of view had therefore ceased; but it is also known that Khons was a late addition to the Theban triad, and I shall subsequently give evidence that the worship was introduced from the south, where it had been conducted when the condition of utility held. The time of introduction to Thebes was the beginning of the eighteenth dynasty, when the priests wished to increase their power by conciliating all worships; and we now see that with their local sun-god Amen-Rā and the goddess Amen-t, with the Northern Mut (Isis) and the Southern Khons, the Theban triad represented the worship of Central, Northern and Southern Egypt.

It is an important fact to bear in mind that in the North of Egypt in early times the stellar temples were more particularly directed to the north, while south of Thebes, so far as I know, there is only one temple so directed. It is suggested, therefore, that the Theban priests amalgamated the northern and southern cults, probably for political purposes. There is evidence that the priests were at heart more sympathetic with the southern cults, and a further investigation of this matter may eventually help us in several points of Egyptian history.

It will have been noticed also that so far as we have gone, whether discussing solar or stellar temples, we have had to associate the cults carried on in most of them with some particular season of the year. If I am right, in the worships at Denderah, Medînet-Habû, and Karnak, we have a strict reference to the year, and in Egypt the year was always, as it is now, associated with the rise of the river.

The sacred river must now occupy our attention for awhile; we must become familiar with its phenomena, and the divisions of time and the calendar systems which were associated with them.

CHAPTER XXIII.

THE EGYPTIAN YEAR AND THE NILE.

Our researches so far leave no doubt upon the question that a large part of the astronomical activity of the earliest Egyptians had reference to observations connected especially with New Year's Day. It has been made abundantly clear, too, that in very early times the Egyptians had a solar year commencing at the Summer Solstice, and that this solstice was then, and is now, coincident with the arrival of the Nile flood at Heliopolis and Memphis, the most important centres of northern Egyptian life during the early dynasties.

In the dawn of civilisation it was not at all a matter of course that the sun should be taken as the measurer of time, as it is now with us; and in this connection it is worth while to note how very diverse the treatment of this subject was among the early peoples. Thus, for instance, it was different in Egypt from what it was in Chaldæa and Babylonia, and later among the Jews. In the Egyptian inscriptions we find references to the moon, but they prove that she occupied quite a subordinate position to the sun, at least in the later times. The week of seven days was utterly unknown amongst the Egyptians. Everything that can be brought forward in its favour belongs to the latest periods. The passage quoted by Lepsius from the Book of the Dead proves nothing, since, according to Krall, an error has crept into his translation. In Babylonia it would seem that the moon was worshipped as well as the sun; and it was thus naturally used for measuring time; and, so far as months were concerned, this, of course, was quite right. In Babylonia, too, where much desert travel had to be undertaken at night, the movements of the moon would be naturally watched with great care.

An interesting point connected with this is that, among these ancient peoples, the celestial bodies which gave them the unit period of time by which they reckoned were practically looked upon in the same category. Thus, for instance, in Egypt the sun being used, the unit of time was a year; but in Babylonia the unit of time was a month, for the reason that the standard of time was the moon. Hence, when periods of time were in question, it was quite easy for one nation to conceive that the period of time used in another was a year when really it was a month, and *vice versâ*. It has been suggested that the years of Methuselah and other persons who are stated to have lived a considerable number of years were not solar years but lunar years—that is, properly, lunar months. This is reasonable, since, if we divide the numbers by twelve, we find that they come out very much the same length as lives are in the present day, and there is no reason why this should not be so.

There seems little doubt that the country in which the sun was definitely accepted as the most accurate measurer of time was Egypt.

Rā, the sun, was the chief god of ancient Egypt. He was worshipped throughout the various nomes. Even the oldest texts (*cf.* that of Menkaurā in the British Museum) tell of the brilliant course of Rā across the celestial vault and his daily struggle with darkness.

"The Egyptians," says Ranke in the first chapter of his "Universal History," which is devoted to Egypt, "have determined the motion of the sun as seen on earth, and according to this the year was divided, in comparison with

Babylon, in a scientific and practically useful way, so that Julius Cæsar adopted the calendar from the Egyptians and introduced it into the Roman Empire. The other nations followed suit, and since then it has been in general use for seventeen centuries. The calendar may be considered as the noblest relic of the most ancient times which has influenced the world."

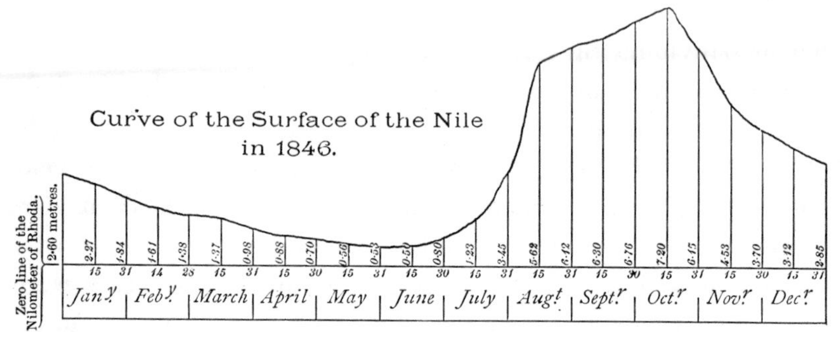

THE ANNUAL RISE AND FALL OF THE NILE. (*From Horner.*)

Wherever the ancient Egyptians came from—whether from a region where the moon was the time-measurer or not—so soon as they settled in the valley where the Nile then, as now, like a pendulum slowly beat the years by its annual inundation at the Summer Solstice, the solar basis of their calendar was settled. Hence it was Nature, the Nile—on the regulation of which depended the welfare of the country—which facilitated the establishment of the Egyptian year. Solstice and Nile-flood are the turning-points of the old Egyptian year.

That Egypt is the gift of the Nile is a remark we owe to the Father of History, who referred not only to the fertilising influence of the stream, but to the fact that the presence of the Nile, and its phenomena, are the conditions upon which the habitability of Egypt altogether depends. That the Egyptian year and that part of Egyptian archæology and myth which chiefly interests astronomers are also the gift of the Nile, is equally true.

The heliacal rising of Sirius and other stars at the time of the commencement of the inundation each year; all the myths which grew out of the various symbols of the stars so used; are so many evidences of the large share the river, with its various water-levels at different times, had in the national life. It was, in fact, the true and unique basis of the national life.

In this the Nile had a compeer or even compeers. What the Nile was to Egypt the Euphrates and Tigris were to a large region of Western Asia, where also we find the annual flood a source of fertility, a spectacle which inspired poets, and an event with which astronomers largely occupied themselves.

What more natural than that Euphrates, Tigris, and Nile were looked upon as deities; that the gods of the Nile valley on the one hand, and of the region watered by the Euphrates and Tigris, on the other, were gods to swear by; that they were worshipped in order that their benign influences might be secured, and that they had their local shrines and special cults?

HĀPI, THE GOD OF THE NILE.

The god sacred to the Euphrates and Tigris was called Ea.

The god sacred to the Nile was called Hāpi. The name is the same as that of the bull Apis, the worship of which was attributed to Mena.[1] Certainly Mena, Mini, or Menes, as he is variously called, was fully justified in founding the cult of the river-god, for he first among men appears to have had just ideas of irrigation, and I have heard the distinguished officers who have lately been responsible for the irrigation system of to-day speaking with admiration of the ideas and works of Mena. Whether the Tigris had a Mena in an equally early time is a point on which history is silent; but, according to the accounts of travellers, the Tigris in flood is even more majestic than the Nile, and yet the latter river in flood is a sight to see—a whole fertile plain turned, as it were, into an arm of the sea, with here and there an island, which, on inspection, turns out to be a village, the mud houses of which too often are undermined by the lapping of the waves in the strong north wind.

There is no doubt that the dates of the rise of these rivers not only influenced the national life, but even the religions of the dwellers on their banks. The Euphrates and Tigris rise at the Spring Equinox—the religion was equinoxial, the temples were directed to the east. The Nile rises at a solstice—the religion was solstitial and the solar temples were directed no longer to the east. To the Egyptians the coming of the river to the parched land was as the sunrise chasing the darkness of the night; the sun-god of day conquering the star-gods of night; or again the victorious king of the land slaughtering his enemies.

Egypt, in the words of Amru, first appears like a dusty plain, then as a fresh sea, and finally as a bed of flowers.

It might be imagined at first sight that as the year was thus determined, so to speak, by natural local causes, the divisions or seasons would be the same as those which Nature

[1] Maspero, "Hist. Anc." xi. 10.

has given *us*. This is not so. The river and land conditions are so widely different.

By no one, perhaps, have the actual facts been so truly and poetically described as by Osborn, who thus pictures the low Nile[1] :—

"The Nile has shrunk within its banks until its stream is contracted to half its ordinary dimensions, and its turbid, slimy, stagnant waters scarcely seem to flow in any direction. Broad flats or steep banks of black, sun-baked Nile mud, form both the shores of the river. All beyond them is sand and sterility; for the hamseen, or sand-wind of fifty days' duration, has scarcely yet ceased to blow. The trunks and branches of trees may be seen here and there through the dusty, hazy, burning atmosphere, but so entirely are their leaves coated with dust that at a distance they are not distinguishable from the desert sand that surrounds them. It is only by the most painful and laborious operation of watering that any tint approximating to greenness can be preserved at this season even in the pleasure-gardens of the Pacha. The first symptom of the termination of this most terrible season is the rising of the north wind (the Etesian wind of the Greeks), blowing briskly, often fiercely, during the whole of the day. The foliage of the groves that cover Lower Egypt is soon disencumbered by it of the dust, and resumes its verdure. The fierce fervours of the sun, then at its highest ascension, are also most seasonably mitigated by the same powerful agency, which prevails for this and the three following months throughout the entire land of Egypt."

Then comes the inundation :—

"Perhaps there is not in Nature a more exhilarating sight, or one more strongly exciting to confidence in God, than the rise of the Nile. Day by day and night by night, its turbid tide sweeps onward majestically over the parched sands of the waste, howling wilderness. Almost hourly, as we slowly ascended it before the Etesian wind, we heard the thundering fall of some mud-bank, and saw, by the rush of all animated Nature to the spot, that the Nile had overleapt another obstruction, and that its bounding waters were diffusing life and joy through another desert. There are few impressions I ever received upon the remembrance of which I dwell with more pleasure than that of seeing the first burst of the Nile into one of the great channels of its annual overflow. All Nature shouts for joy. The men, the children, the buffaloes, gambol in its refreshing waters, the broad waves sparkle with shoals of fish, and fowl of every wing flutter over them in clouds. Nor is this jubilee of Nature confined to the higher orders of creation. The moment the sand becomes moistened by the

[1] "Monumental Egypt," chapter i.

approach of the fertilising waters, it is literally alive with insects innumerable. It is impossible to stand by the side of one of these noble streams, to see it every moment sweeping away some obstruction to its majestic course, and widening as it flows, without feeling the heart to expand with love and joy and confidence in the great Author of this annual miracle of mercy."

DIFFERENT FORMS OF THOTH.

After the flood comes the sowing time. The effects of the inundation, as Osborn shows in another place,

"exhibit themselves in a scene of fertility and beauty such as will scarcely be found in another country at any season of the year. The vivid green of the springing corn, the groves of pomegranate-trees ablaze with the rich scarlet of their blossoms, the fresh breeze laden with the perfumes of gardens of roses and orange thickets, every tree and every shrub covered with sweet-scented flowers. These are a few of the natural beauties that welcome the stranger to the land of Ham. There is considerable sameness in them, it is true, for he would observe little variety in the trees and plants, whether he first entered Egypt by the gardens of Alexandria or the plain of Assouan. Yet is it the same everywhere, only because it would be impossible to make any addition to the sweetness of the odours, the brilliancy of the colours, or the exquisite beauty of the many forms of vegetable life, in the midst of which he wanders. It is monotonous, but it is the monotony of Paradise."

The flood reaches Cairo on a day closely approximating to that of the Summer Solstice. It attains its greatest height, and begins to decline near the Autumnal Equinox. By the Winter Solstice the Nile has again subsided within its banks and resumed its blue colour. Seed-time has occurred in this interval.

Beginning with the inundation (Summer Solstice) we have—

(1) The season or *tetramene* of the inundation, July—October.
(2) " " " sowing, November—February.
(3) " " " harvest, March—June.

From the earliest times the year was divided into twelve months, as follows, the leading month being dedicated to the God of Wisdom, Thoth (Tehuti):—

Inundation ...	Thoth	End of June (Gregorian).
	Phaophi	" July.
	Athyr	" August.
	Choiak	" September.
Seed-time ...	Tybi	" October.
	Menchir	" November.
	Phamenoth	" December.
	Pharmouthi	" January.
Harvest ...	Pachons	" February.
	Payni	" March.
	Epiphi	" April.
	Mesori	" May.

The terms for the seasons and months are found even on the building material of the largest pyramid of Dashûr, and in the oldest records we already find calendar indications. On the steles of the Mastăbas, in which the deceased prays Anubis for a good sepulture, we find a list of the festal days on which sacrifices are to be offered for the dead.

A modern calendar (given both by Brugsch and De Rougé) is, doubtless, a survival from old Egyptian times. It is good for the neighbourhood of Cairo, and the relation of the important days of the inundation to the solstice, in that part of the river, is as follows:—

Night of the drop	11 Payni	
	15 ,,	Summer solstice.
Beginning of the inundation	18 ,,	3 days after.
Assembly at the nilometer	25 ,,	10 ,,
Proclamation of the inundation	26 ,,	11 ,,
Marriage of the Nile	18 Mesori	63 ,,
The Nile ceases to rise	16 Thoth	96 ,,
Opening of the dams	17 ,,	97 ,,
End of the greater inundation	7 Phaophi	117 ,,

In order to show how the astronomy of the ancient Egyptians—to deal specially with them—was to a large extent concerned with the annual flood and all that depended upon that flood; and how the first tropical year used on this planet, so far as we know, was established, it is important to study the actual facts of the rise somewhat closely, not only for Egypt generally, but for several points in the line, some thousand miles in extent, along which in the earliest times cities and shrines were dotted here and there.

Time out of mind the fluctuations in the height of the river have been carefully recorded at different points along the river. In the "Description de l'Égypte" we find a full description of the so-called nilometer at Aswân (First Cataract), which dates from a remote period, perhaps as early as the fifth dynasty.

In Ebers' delightful book on Egypt space is given to the description of the much more modern one located at Rôda.

The nilometer, or "mikyās," on the island of Rôda, now

visible, is stated to have replaced one which was brought thither from Memphis at some unrecorded date. Makrīzī in 1417, according to Ebers, saw the remains of the older nilometer.

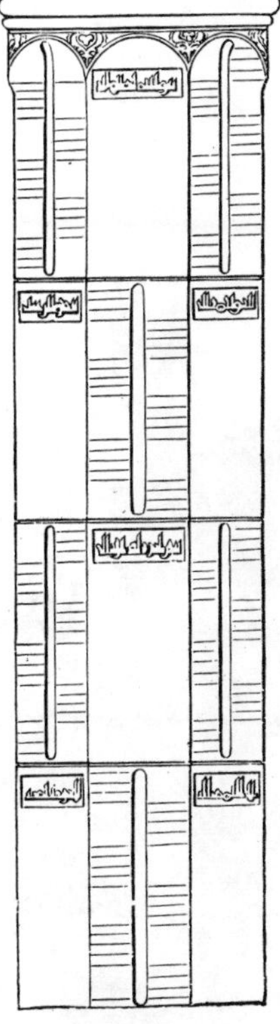

SCALE OF THE NILOMETER AT RÔDA.

The present mikyās is within a covered vault or chamber, the roof being supported on simple wooden pillars. In a quadrangular tank in communication with the river by a canal is an octagon pillar on which the Arabic measurements are inscribed. These consist of the pic (variously called ell or cubit) = 0·54 metre, which is divided into twenty-four kirats. In consequence of the rise of the river bed in relatively recent times, the nilometer is submerged at high Nile to a depth of two cubits.

The rise of the Nile can now be carefully studied, as gauges are distributed along the river. We have the Aswân gauge from 1869, the Armant gauge from 1887, the Suhag gauge from 1889, and the Asyût gauge from 1892. The distances of these gauges from Aswân are as follows:—

	Kilometres.
Aswân	0
Armant	200
Suhag	447
Asyût	550
Rôda	941

The Rôda gauge is not to be depended on, as the movements of the barrage regulation destroy its value as a record.

The heights of these gauges above mean sea-level are as follows:—

	Metres.
Aswân	84·158
Armant	69·535
Suhag	56·00
Asyût	53·10
Rôda	13·14

THE ISLAND OF RÔDA.

Great vagueness arises in there being no very obvious distinction between the gauge readings reached in summer and that from which the rise is continuous. There are apparently rainfalls in the end of spring of sufficient power to raise the Nile visibly in summer, just as muddy rises have been seen in winter to pass down the valley, leaving a muddy mark on the rocks at Aswân and Manfalūt. Independently of the actual gauge-reading of the rise, there are facts about it which strike every beholder. At the commencement of the rise we have the *green water*. This occurs in June, but varies in date as much as the top of the flood varies.

From the fact that modern observations show that the very beginning of the rise, and the first flush, second flush and final retirement vary, it seems evident that the ancient Egyptians could not have had any fixed zero-gauge or time for the real physical fact of the rise, but must have deduced from a series of observations either a mean period of commencement, or a mean arrival of the red water, or a mean rising up to a certain gauge.

First, to deal with the green water. Generally when the rise of an inch or two is reported from the nilometer at Rôda, the waters lose the little of clearness and freshness they still possessed. The green colour is the slimy, lustreless hue of brackish water within the tropics, and no filter that has yet been discovered can render such water clear. The colour is really due to algæ.

Happily, the continuance of this state of the water seldom exceeds three or four days. The sufferings of those who are compelled to drink it in this state, from vesical disease, even in this short interval, are very severe. The inhabitants of the cities generally provide against it by Nile water stored in reservoirs and tanks.

Colonel Ross, R.E., noticed in 1887 and in 1890, when, owing to the slow retreat of the Nile, the irrigation officers had to hold back many basins in the Gîzeh province, and also in 1888 when the water remained long stagnant, that the basin water got green—showed the algæ and smelt marshy—just as the June green water does.

Hence it has been argued that, as the Nile water in the bed of the stream—even in very slow-flowing back-waters—does not become green, the greenness must be produced by an almost absolute stagnation of the water. We know of great marshes up above Gondokoro, and hence it is thought that the green water of summer, which comes on suddenly, is this marsh-water being pushed out by the new water from behind, and that is why it heralds the rise. No one has so far minutely observed the gradual intrusion of the green water.

The rise of the river proceeds rapidly, and the water gradually becomes more turbid. Ten or twelve days, however, elapse before the development of the last and most extraordinary of all the appearances of the Nile, thus described by Mr. Osborn[1] :—

"It was at the end of—to my own sensations—a long and very sultry night, that I raised myself from the sofa upon which I had in vain been endeavouring to sleep, on the deck of a Nile boat that lay becalmed off Benisoueff, a town of Middle Egypt. The sun was just showing the upper limb of his disc over the eastern mountains. I was surprised to see that when his rays fell upon the water a deep ruddy reflection was given back. The depth of the tint increased continually as a larger portion of his light fell upon the water, and before he had entirely cleared the top of the hill it presented the perfect appearance of a river of blood. Suspecting some delusion, I rose up hastily, and, looking over the side of the boat, saw there the confirmation of my first impression. The entire body of the water was opaque and of a deep red colour, bearing a closer resemblance to blood than to any other natural production to which it could be compared. I now perceived that during the night the river had visibly risen several inches. While I was gazing at this great sight the Arabs came round me to explain that it was the Red Nile. The redness and opacity of the water, in this extraordinary condition of the river, are subject to constant variations. On some days, when the rise of the river has not exceeded an inch or two, its waters return to a state of semi-transparency, though during the entire period of the high Nile they never lose the deep red tinge which cannot be separated from them. It is not, however, like the green admixture, at all deleterious; the Nile water is never more wholesome or more deliciously refreshing than during the overflow. There are other days when the rise of the river is much more rapid, and then the quantity of mud that is suspended in the water exceeds, in

[1] "Monumental Egypt," chapter i.

Upper Egypt, that which I have seen in any other river. On more than one occasion I could perceive that it visibly interfered with the flow of the stream. A glassful of it in this state was allowed to remain still for a short time. The upper portion of it was perfectly opaque and the colour of blood. A sediment of black mud occupied about one-quarter of the glass. A considerable portion of this is deposited before the river reaches Middle and Lower Egypt. I never observed the Nile water in this condition there, and indeed no consecutive observations exist of the reddening of the water. It is quite clear that the reddening cannot come from the White Nile, but must be the first floods of the Blue Nile and Bahral Azral coming down."

One of the most important matters for the purposes of our present inquiry is connected with the influence upon local calendars, in different parts of the Nile valley, of the variations of the phenomena upon which the Egyptians depended for the marking of New Year's Day.

If the *solstice* had been taken alone, the date of it would have been the same for all parts of the valley; but certainly the solstice was not taken alone, and for the obvious reason, that they wanted something to warn them of the Nile rise, and in the lower reaches of the river the rise precedes the solstice. Nor was the heliacal rising of Sirius, of which more presently, taken alone.

But it was chiefly a question of the arrival of the Nile flood, and the date of the commencement of the Nile flood was by no means common to all parts of Egypt.

Now it is to be gathered from the modern gauges that it takes the flood some time, as we can easily imagine, to pass down the 600 miles between Elephantine and Cairo.

In the early flood, rising from, say, one cubit Aswân to six cubits, where there are many dry sandbanks, and the spreading out of the river is considerable, and there is an absence of overlapping flushes from behind, the rate goes up to fifteen days, and the *earliest* indication of the rise may take longer still, but this is very difficult to observe.

The rate in flood is $1\frac{3}{4}$ days from Wādy Halfa to Aswân, and six days from Aswân to Rôda (941 kilometres). In very high Niles this is perhaps accelerated to five days.

There is, therefore, a very great difference in time and rate between Green and Red Nile.

The rise is 45ft. at Aswân, 38 at Thebes, and 25 at Cairo.

From the data obtained at the gauges named, which have been kindly forwarded to me by Mr. Garstin, the Under Secretary of State of the Public Works Department of Egypt, I have ascertained that the average time taken by the first indication of the flood to travel between Thebes and Memphis is now about nine days.

It must be remembered, however, that the river-bed is now higher than formerly; the land around Thebes, according to Budge, has been raised about nine feet in the last 1,700 years.

If, therefore, at each great city, such as Thebes and Heliopolis, New Year's Day depended absolutely on the arrival of the inundation, not only would the day have been uncertain, but the difference of time in the arrival of the flood at various places along the river would represent a difference in the New Year's Days of those places, compared to which our modern differences of local time sink into insignificance, for they only touch hours of the day.

The great difficulty experienced in understanding the statements generally made concerning the Nile-rise is due to the fact that the maximum flood is, as a rule, registered in Cairo upwards of forty days after the maximum at Aswân.

For the following account of how this is brought about I am indebted to the kindness of Colonel Ross, R.E.:—

"The behaviour of the flood at the Aswân gauge is as follows: Between August 20 and 30 a good average gauge of 16 cubits is often reached, and between August 27 and September 3 there is often a drop of about

30 centimetres. The August rise is supposed to be mostly due to the Blue Nile and Atbara River. Between September 1 and 8 the irrigation officers generally look for a maximum flood-gauge of the year at Aswân. This is supposed to be the first flush of the White Nile. In the middle of September there are generally two small flushes, but the last twenty days of September are generally distinctly lower than that of the first week. The final flush of the Nile is seldom later than the 21st to 25th September.

"All this water does not merely go down the Nile; it floods the different basins. The opening of these basins begins from the south to the north. This operation is generally performed between the 29th September and the 22nd October. The great Central Egypt basins are not connected with the Nile for purposes of discharge into the river between Asyût and near Wasta, or a distance of 395 − 90 kilometres = 305 kil.

"The country in the middle or Central Egypt is broad, and thus there is an enormous quantity of water poured out of these basins into the lower reaches of the river about the 20th October, which seriously raises the Nile at Cairo, and in a good average year will bring the Cairo gauge (at Rôda) up to the maximum of the year on or about October 22, and hence it is that the guide-books say the Nile is at its highest in the end of October.

"A gauge of 16½ cubits at Aswân while the basins are being filled does not give more than 21 cubits at Rôda (Cairo), but, as the basins with a 16½ gauge will fill by the 10th September, it follows that a 16½ to 16 cubit gauge at Aswân will not give a constant Cairo gauge, as the great mass of water passes by the basins and reaches Cairo. Hence we have frequently the paradox of a steady or falling gauge at Aswân showing a steady rise at Cairo.

"If the gauge at Aswân keeps above 16 cubits to near the end of September, the basin-emptying is much retarded, as the emptying at each successive basin fills the Nile above the 16 cubit level; *hence the lower halves of the basins do not flow off*, and thus, when the great Middle Egypt basins are discharged, they do not raise the Nile so much as they do when the last half of September Nile is below 16 at Aswân.

"In years like 1887 and 1892, which differ from each other only in date of maximum gauge at Aswân, the river, having filled the basins in fifteen to twenty days instead of in twenty-five to thirty days, comes down to Cairo in so largely increased a volume that a really dangerous gauge of 25 cubits at Cairo is maintained for over a fortnight (the average October gauge in Cairo is about 23 cubits), and from September 10 to October 25 the river remains from 24 cubits to 25½ cubits, and the Middle Egypt basins discharge so slowly that the opening day is hardly traceable on the Cairo gauge.

"In the 1878 flood, which was the most disastrous flood possible, the river rose in the most abnormal fashion, and on October 3 attained 18 cubits at Aswân. This breached the Delta, and in addition so delayed the Upper Egypt basins emptying, from the reason before given, that the wheat was sown too late, and got badly scorched by the hot winds of March and April."[1]

[1] The modern Egyptians still hold to the old months for irrigation. 7 Taba = January 15, is commencement of wheat irrigation; 30 Misra is the last safe date for sowing maize in the Delta; 1st Tut is the date of regulating the bridges = September 8 in Upper Egypt.

CHAPTER XXIV.

THE YEARS OF 360 AND 365 DAYS.

WHETHER the Egyptians brought their year with them or invented it in the Nile valley, there is a belief that it at first consisted of 360 days only, that is, $5\frac{1}{4}$ days too little. It is more likely that they brought the lunar month with them, taking it roughly as 30 days ($30 \times 12 = 360$), than that they began with such an erroneous notion of the true length of the solar year, seeing that in Egypt, above all countries in the world, owing to the regularity of the inundation, the true length could have been so easily determined, so soon as that regularity was recognised. We must not in these questions forget to put ourselves in the place of these pioneers of astronomy and civilisation; if we do this, we shall soon see how many difficulties were involved in determining the true length of such a cycle as a year, when not only modern appliances, but all just ideas too, were of necessity lacking.

Since 360 days do not represent the true length of the year, it is clear that any nation which uses such a year as that will find the seasons and festivals sweeping through the year. Further, such a year is absolutely useless for the agriculturist or the gardener, because after a time the same month, to say nothing of the same day of the month, will not mean reaping-time, will not mean sowing-time, or anything else.

Still, it is right that I should state that all authorities are not agreed as to the use of this year of 360 days; at all events, during the times within our ken. Maspero[1] states:—

[1] "Histoire ancienne des Peuples de l'Orient," p. 72.

"Des observations nouvelles, faites sur le cours du soleil, décidèrent les astronomes à intercaler chaque année, après le douzième mois, et avant le premier jour de l'année suivante, cinq jours complémentaires, qu'on nomma *les cinq jours en sus de l'année* ou jours *epagomènes* (*epacts*). L'époque de ce changement était si ancienne que nous ne saurions lui assigner aucune date, et que les Egyptiens eux-mêmes l'avaient reportée jusque dans les temps mythiques antérieurs a l'avènement de Mini."

Ideler[1] is of the same opinion as Maspero :—

"I do not hesitate . . . . . . to declare that the existence of such a time cycle —used without reference to the course of the sun or moon simply for the sake of simple figures—is extremely doubtful to me."

Krall remarks (p. 17) :—

"It is probable that the year of 360 days dates from the time before the immigration into the Nile valley, when the Egyptians were unguided by the regular recurrence of the Nile flood. In any case, this must soon have convinced the priests that the 360-days year did not agree with the facts. But it is well known to everybody familiar with these things how long a period may be required before such determinations are practically realised, especially with a people so conservative of ancient usages as the Egyptians."

And on this ground, apparently, he joins issue with the authorities already quoted :—

"The Egyptian monuments have contradicted Ideler in this respect. The trilingual inscription of Tanis testifies expressly that it has only 'later become usual to add the five epagomenes;' that, therefore, the year originally had 360 days, which were divided into twelve months of thirty days each."

Krall also argues that the expressions great and little year and their hieroglyphics referred to the years of 365 and 360 days respectively, and adds :—

"If we inquire into the time at which the epagomenes were introduced, we can only fix approximate dates. If the calendars of the Mastabas, complete as they are, do not mention the epagomenes, whereas inscriptions of the period of the Amenamhāts refer to them, this can only be due to the circumstance that the epagomenes were only introduced in the meantime, but probably nearer the upper than the lower limit. . . . For the sake of completeness, we may mention that, according to Censorinus, the five epagomenes were introduced by the King Arminon. . . . . Louth conjectures that Arminon is identical with Amenamhāt I.,

[1] "Chronologie," i., p. 70.

under whom the epagomenes are first met with. But since, between Nitokris and Amenamhāt I., there is a period of 500 years void of records, and the name Arminon has nothing to do with Amenamhāt, we can hardly share this view."

However this knotty point may subsequently be settled by Egyptologists, from the astronomer's point of view the words of Ideler[1]—"Had ignorance lead to the establishment of a year of 360 days, yet experience would have led to its rejection in a few years"—will carry conviction with them. Indeed, one may ask whether it is not possible that the use of the 360-day year, and the complications which it involved, may have had something to do with the foundation of the solar temples.

Let us attempt to put ourselves, in imagination, in the place of the ancient Egyptians after the use of this 360-day year had been continued for any length of time. It is perfectly certain that now in this part of the Nile valley, now in that, everybody, from Pharaoh to fellah, must have got his calendar into the most hopeless confusion, compared with which "the year of confusion" was mere child's-play, and that the exact determination of the times, either of state functions or sowing, reaping, or the like, by means of such a calendar would have been next to impossible.

As each year dropped $5\frac{1}{4}$ days, it is evident that in about seventy years $(\frac{365 \cdot 25}{5 \cdot 25})$ a cycle was accomplished, in which New Year's Day swept through all the months. The same month (so far as its name was concerned) was now in the inundation time, now in the sowing time, and so on. Of fixed agricultural work for such months as these there could be none.

It must have been, then, that there were local attempts to retain the coincidences between the true and the calendar year—intercalation of days or even of months being introduced, now in one place, now in another; and these attempts, of course,

[1] *Op. cit.*, p. 187.

would make confusion worse confounded, as the months might vary with the district, and not with the time of year.

That this is what really happened is, no doubt, the origin of the stringent oath required of the Pharaohs in after times, to which I shall subsequently refer.

To acknowledge that the calendar year was wrong implied that they knew the length of the true one. How had they found it out? I think there can be no question that this knowledge had come to them by observations either of the solstices or the equinoxes. It is true they had the inundation; but, as we have seen, the rise is not absolutely regular, and the inundation takes many days to travel from Philæ to Cairo (Memphis). If, then, the inundation had fixed the beginning of the year, each nome would have its special New Year's Day, and this would never have been tolerated by a settled government embracing the whole Nile valley, especially as each king's reign was supposed to commence on New Year's Day.

It seems, then, that the solstitial temples and the pyramids were, if not actually requisite for settling the matter, at all events all that was necessary, if they existed.

But now comes in a most interesting and important point. If observations of the sun at solstice or equinox had been alone made use of, the true length of the year would have been determined in a few years. But the next scene in Egyptian history shows us that the true length of the year was not determined, but only an approximation to it.

How was this? The astronomical answer is very simple.

I have already referred to the common practice of all ancient peoples that we know of to make sacrifices at dawn, and have shown how, in order to do this, they took their time from a star rising before the sun. An observation of the so-called "heliacal rising" of a star—if the star were properly chosen—would give them the interval necessary for their preparations before the sun itself appeared; and, as the highest festival of all was that of New Year's Day, it was especially important that the work should be well done then.

Now, if the stars had no precessional movement, the sun and stars, after each interval of a true year, would be in exactly the same position; but in consequence of the stars having the precessional movement to which I have before referred, the star so observed and the sun will *not* be in exactly the same position after the interval of a true year. On this account, then, the difference of time between the heliacal risings will not represent the length of a true year. But, further, the heliacal rising of the star will not take place on the same day for the whole of Egypt, the difference between Thebes and Memphis, depending upon their latitudes, amounting to about four days; and, further still, the almost constant mists in the mornings in the Nile valley prevent accurate observations of the moment of rising.

Still, as a matter of fact, the Egyptians defined their new year by the rising of a star, and the length of it by the interval separating two heliacal risings. Such a year could not be accurate; and again, as a matter of fact, their correction was not accurate, for the year was defined now as consisting of 365 days. It seems clear from this that the correction was made before the solar temples were in use.

In any case the year of 360 days had naturally to give way, and it ultimately did so, in favour of one of 365. The precise date of the change is, as we have seen, not known.[1] The five days were added as epacts or epagomena; the original months were not altered, but a "little month" of five

[1] Krall, *loc. cit.*, p. 20.

days was interpolated at the end of the year between Mesori of one year and Thoth of the next, as already stated.

When the year of 365 days was established, it was evidently imagined that finality had been reached; and, mindful of the confusion which, as we have shown, must have resulted from the attempt to keep up a year of 360 days by intercalations, each Egyptian king, on his accession to the throne, bound himself by oath before the priest of Isis, in the temple of Ptah at Memphis, not to intercalate either days or months, but to retain the year of 365 days as established by the Antiqui.[1] The text of the Latin translation preserved by Nigidius Figulus cannot be accurately restored; only thus much can be seen with certainty.

To retain this year of 365 days, then, became the first law for the king, and, indeed, the Pharaohs thenceforth throughout the whole course of Egyptian history adhered to it, in spite of their being subsequently convinced, as we shall see, of its inadequacy. It was a Macedonian king who later made an attempt to replace it by a better one.

We may reckon upon the conservatism of the priests of the temples retaining the tradition of the old rejected year in every case. Thus even at Philæ in late times, in the temple of Osiris, there were 360 bowls for sacrifice, which were filled daily with milk by a specified rotation of priests. At Acanthus there was a perforated cask into which one of the 360 priests poured water from the Nile daily.

Indeed, these temple ceremonials are an evidence of their antiquity, and the further we put back the change from the 360 to 365 days, the greater the antiquity we must assign to them, and therefore to the temples themselves.

[1] Mommsen, "Chronologie," p. 258.

CHAPTER XXV.

THE VAGUE AND THE SIRIAN YEARS.

During three thousand years of Egyptian history the beginning of the year was marked by the rising of Sirius, which rising took place nearly coincidently with the rise of the Nile and the Summer Solstice.

I have insisted upon the regularity of the rise of the Nile affording the ancient Egyptians, so soon as this regularity had been established, a moderately good way of determining the length of the year, but we have seen they did not so employ it.

It is also clear that so soon as the greatest northing and southing of the sun rising or setting at the solstices had been recognised, and the intervals between them in days had been counted, a still more accurate way would be open to them. The solstice *must* have occurred with greater regularity than the rise of the river, so that as accuracy of definition became more necessary the solstice would be preferred. The solstice was common to all Egypt; the commencement of the inundation was later as the place of observation was nearer the mouth of the river. This means they also did not employ, at all events in the first instance. Of the three coincident, or nearly coincident, phenomena, the rise of the Nile, the Summer Solstice, and the rising of Sirius, they at first chose the last.

According to Biot the heliacal rising of Sirius *at the solstice* took place on July 20 (Julian), in the year 3285 B.C.; and according to Oppolzer it took place on July 18 (Julian), in the year 3000 B.C.

But this is too general a statement, and it must be modified here. There was a difference of seven days in the date of the heliacal rising, according to the latitude, from southern Elephantine and Philæ, where the heliacal rising at the solstice was noted first, to northern Bubastis. There was a difference of four days between Memphis and Thebes, so that the connection between the heliacal rising and the solstice depended simply upon the latitude of the place. The further south, the earlier the coincidence occurred.

Here we have an *astronomical* reason for the variation in the date of New Year's Day.

There no doubt was a time when the Egyptian astronomer-priests imagined that, by the introduction of the 365-days year, marking its commencement, as I have said, by the rising of one of the host of heaven, they had achieved finality. But, alas, the dream must soon have vanished.

Even with this period of 365 days, the true length of the year had not been reached; and soon, whether by observations of the beginning of the inundation, or by observations of the solstice in some of the solar temples when these had been built, it was found that there was a difference of a day every four years between the beginning of the natural and of the newly-established year, arising, of course, from the fact that the true year is 365 days *and a quarter of a day* (roughly) in length.

With perfectly orientated temples they must have soon found that their festival at the Summer Solstice—which festival is known all over the world to-day—did not fall precisely on the day of the New Year, because, if 365 days had exactly measured the year, that flash of bright sunlight would have fallen into the sanctuary just as it did 365 days before. But what they must have found was that, after an interval of four years, it did not fall on the first day of the month, but on the day following it.

The true year and the newly-established year of 365 days, then, behaved to each other as shown in the following diagram, when the solstice, representing the beginning of the calendar year, occurred on the 1st Thoth of the newly-established calendar year. We should have, in the subsequent years, the state of things shown in the diagram. The solstice would year

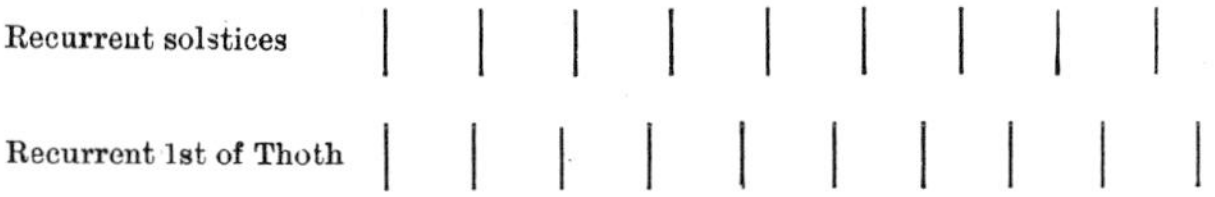

Recurrent solstices | | | | | | | | |

Recurrent 1st of Thoth | | | | | | | | | |

by year occur *later* in relation to the 1st of Thoth. The 1st of Thoth would occur *earlier*, in relation to the solstice; so that in relation to the established year the solstice would sweep forwards among the days: in relation to the true year the 1st of Thoth would sweep backwards.

Let us call the true natural year a *fixed* year: it is obvious that the months of the 365-day year would be perpetually varying their place in relation to those of the fixed year. Let us, therefore, call the 365-day year a *vague* year.

Now if the fixed year were exactly $365\frac{1}{4}$ days long, it is quite clear that, still to consider the above diagram, the 1st of Thoth in the vague year would again coincide with the solstice in 1,460 years, since in four years the solstice would fall on the 2nd of Thoth, in eight years on the 3rd of Thoth, and so on ($365 \times 4 = 1460$).

But the fixed year is not $365\frac{1}{4}$ days long *exactly*. In the time of Hipparchus 365·25 did not really represent the true length of the solar year; instead of 365·25 we must write 365·242392—that is to say, the real length of the year is a little *less* than $365\frac{1}{4}$ days.

Now the length of the year being a little *less*, of course we should only get a second coincidence of the 1st of Thoth vague with the solstice in a *longer* period than the 1460-years cycle; and, as a matter of fact, 1506 years are required to fit the months into the years with this slightly shortened length of the year. In the case of the solstice and the vague year, then, we have a cycle of 1,506 years.

The variations between the fixed and the vague years were known perhaps for many centuries to the priests alone. They would not allow the established year of 365 days, since called the *vague* year, to be altered, and so strongly did they feel on this point that, as already stated, every king had to swear when he was crowned that he would not alter the year. We can surmise why this was. It gave great power to the priests; they alone could tell on what particular day of what particular month the Nile would rise in each year, because they alone knew in what part of the cycle they were; and, in order to get that knowledge, they had simply to continue going every year into their Holy of Holies one day in the year, as the priests did afterwards in Jerusalem, and watch the little patch of bright sunlight coming into the sanctuary. That would tell them exactly the relation of the true solar solstice to their year; and the exact date of the inundation of the Nile could be predicted by those who could determine observationally the solstice, but by no others.

But now suppose that, instead of the solstice, we take the heliacal rising of Sirius, and compare the successive risings at the solstice with the 1st of Thoth.

But why, it will be asked, should there be any difference in the length of the cycles depending upon successive coincidences of the 1st of Thoth with the solstice and the heliacal rising of Sirius? The reason is that stars change their places, and the star to which they trusted to warn them of the beginning of a new year was, like all stars, subject to the effects brought about by the precession of the equinoxes. Not for long could it continue to rise heliacally either at a solstice or a Nile flood.

Among the most important contributors to the astronomical side of this subject are M. Biot and Professor Oppolzer. It is of the highest importance to bring together the fundamental points which have been made out by their calculations. We have determinate references to the heliacal rising of Sirius, to the 1st of Thoth, to the solstice, and to the rising of the Nile in connection with the Egyptian year; but, so far as I have been able to make out, we find nowhere at present any sharp reference to the importance of their correlation with the times of the *tropical* year at which these various phenomena took place. The question has been complicated by the use by chronologists of the Julian year in such calculations; so the Julian year and the use made of it by chronologists have to be borne in mind. Unfortunately, many side-issues have in this way been raised.

The heliacal rising of Sirius, of course—if in those days a true *tropical* year was being dealt with—would have given us a more or less constant variation in the time of the rising over a long period, *on account of its precessional movement;* and M. Biot and others before him have pointed out that the variation, produced by that movement, in the time of the year at which the heliacal rising took place was almost exactly equal to the error of the *Julian* year as compared with the true tropical or Gregorian one. The Sirius year, like the Julian, was about eleven minutes longer than the true year, so that in 3,000 years we should have a difference of about 23 days. Biot showed by his calculations, using the solar tables extant before those of Leverrier, that from 3200 B.C. to 200 B.C. in the Julian year of the chronologists, Sirius had constantly, in each year, risen heliacally on July 20 Julian = June 20 Gregorian. Oppolzer, more recently, using Leverrier's tables, has made a very slight correction to this, which, however, is practically immaterial for the purposes of a general statement. He shows that in the latitude of Memphis, in 1600 B.C., the heliacal rising took place on July 18·6, while in the year 0 it took place on July 19·7, both Julian dates.

The variation from the true tropical year brought about by the precessional movement of Sirius or any other star, however, can be watched by noting its heliacal rising in relation to any physical phenomenon which marks the true length of the tropical year. Such a phenomenon we have in the solstice and in the rising of the Nile, which, during the whole course of historical time, has been found to rise and fall with constancy in each year, the initial rise of the waters, some little way above Memphis, taking place very nearly at the Summer Solstice.

Again, M. Biot has made a series of calculations from which we learn that the heliacal rising of Sirius AT THE SOLSTICE occurred on July 20 (Julian) in the year 3285 B.C., and that in the year 275 B.C., the *solstice* occurred on June 27 (Julian), while the heliacal *rising of Sirius* took place, as before, on July 20 (Julian), so that in Ptolemaïc times, at Memphis, there was a difference of time of about 24 days between the heliacal rising of Sirius and the solstice, and therefore the beginning of the Nile flood in that part of the river. This, among other things, is shown on the next page.

We learn from the work of Biot and Oppolzer, then, that the precessional movement of the star caused successive heliacal risings of Sirius at the solstice to be separated by almost exactly 365¼ days—that is, by a greater period than the length of the

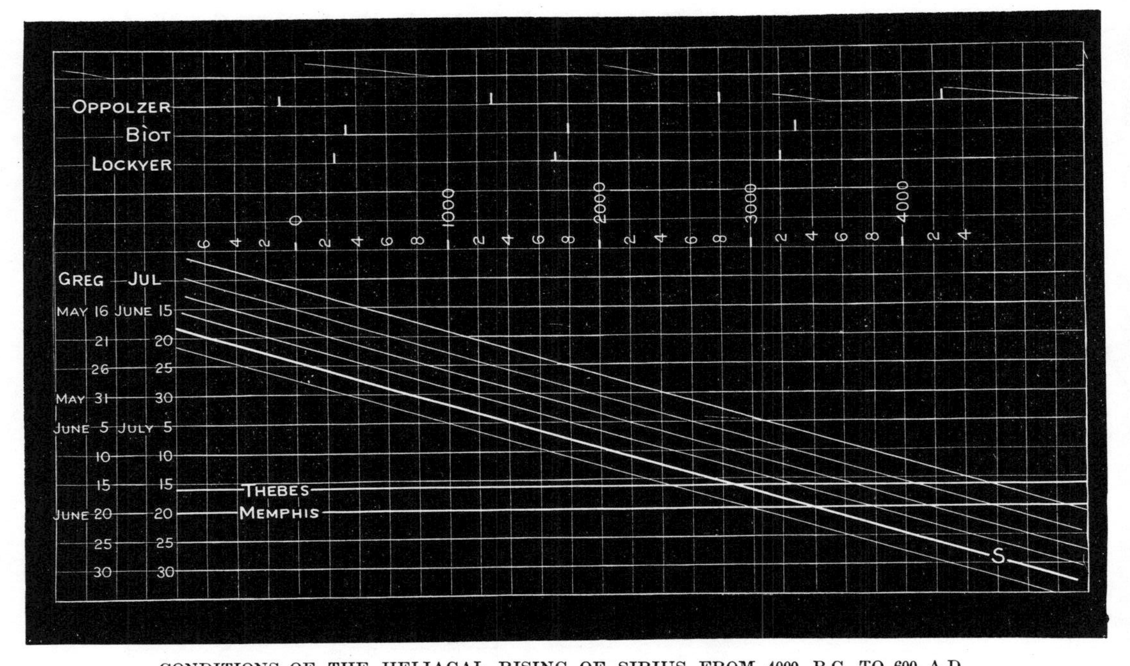

CONDITIONS OF THE HELIACAL RISING OF SIRIUS FROM 4000 B.C. TO 600 A.D.

The diagram shows (1) by white horizontal lines the Gregorian and Julian dates for the rising at Thebes and at Memphis; (2) by the full diagonal line the *Julian* date of the solstice or beginning of the inundation in each century, at a point of the river near Memphis. The fainter lines show the Julian dates for other places where the time of the beginning of the flood differs by three days from the Memphis dates. The interval between each line represents a difference of three days in the arrival of the flood; (3) the interval in days between the heliacal rising and the inundation at different periods and at different points of the river. This can be determined for each century by noticing the interval between the proper diagonal line and that indicating the heliacal rising; (4) by dots at the top of the diagram the commencement of the Sothic period as determined by Oppolzer, Biot, and the author.

true year. So that, in relation to this star, two successive heliacal risings at the 1st of Thoth vague are represented by a period of ($365\frac{1}{4} \times 4 =$) 1461 years, while in the case of the solstices we want 1506.

Now in books on Egyptology the period of 1461 years is termed the Sothic period, and truly so, as it very nearly correctly measures the period elapsing between two heliacal risings at the solstice (or the beginning of the Nile flood) on the 1st of Thoth in the *vague* year.

But it is merely the result of *chance* that $365\frac{1}{4} \times 4$ represents it. It was not then known that the precessional movement of Sirius almost exactly made up the difference between the true length of the year and the assumed length of $365\frac{1}{4}$ days. It has been stated that this period had not any ancient existence, but was calculated back in later times. This seems to me very improbable. I look upon it rather as a true result of observation, the more so as *the period was shortened in later times*, as Oppolzer has shown.

It will be seen that our investigations land us in several astronomical questions of the greatest interest, and that the study is one in which modern computations, with the great accuracy which the work of Leverrier and others gives to them, can come to the rescue, and eke out the scantiness of the ancient records.

To consider the subject further, we must pass from the mere question of the year to that of chronology generally.

CHAPTER XXVI.

THE SOTHIC CYCLE AND THE USE MADE OF IT.

ALTHOUGH it is necessary to enter somewhat into the domain of chronology to really understand the astronomical observations on which the Egyptian year depended and the uses made of the year, I shall limit myself to the more purely astronomical part. To go over the already vast literature is far from my intention, nor is it necessary to attempt to settle all the differences of opinion which exist, and which are so ably referred to by Krall in his masterly analysis,[1] to which I own myself deeply indebted. The tremendously involved state of the problem may be gathered from the fact that the authorities are not yet decided whether many of the dates met with in the inscriptions really belong to a fixed or a vague year!

Let us, rather, put ourselves in the place of the old Egyptians, and inquire how, out of the materials they had at hand, a calendar could be constructed in the simplest way.

They had the vague year and the Sirius year, so related, as we have seen, that the successive coincidences of the 1st Thoth in both years took place after an interval of 1460 years. Now, for calendar purposes, they wanted not only to know the days of the years, but the years of the cycle. This latter is the only point we need consider here. How were they to do this? The *easiest* way would be to conceive a great year or *annus magnus*, consisting of 1460 years, each day of which would represent four years in actual time; and further, to consider everything that happened, which had to be thus chronicled, to take place

[1] "Studien zur Geschichte der Alten Aegypten," I. Wien, 1881.

on the 1st of Thoth in each year. How would this system work? During the first four years, at the beginning of a cycle, the 1st Thoth vague would happen on the 1st Thoth of the cycle. During the next four years the 1st Thoth of the vague year would fall on the fifth epact, and so on; so, as the cycle swept onward, each group of four years would be marked by a date in the cycle, which would allow the place of the group of years in the cycle to be exactly defined. But as the cycle swept onward, the date would sweep backward among the months of the great sacred year until its end.

To make this clear, it will be well to construct another diagram somewhat like the former one.

Let us map out the 1460 years which elapsed between two successive coincidences between the 1st of Thoth in the vague year and the heliacal rising of Sirius at the solstice, so that we can see at a glance the actual number of years from any start-point (= 0) at which the 1st of Thoth in the vague year occurred successively further and further from the heliacal rising, until at length, after a period of 1460 years, it coincided again. As the Sirius-year is longer than the vague one, the first vague year will be completed before the first Sirius-year, hence the second vague year will commence just before the

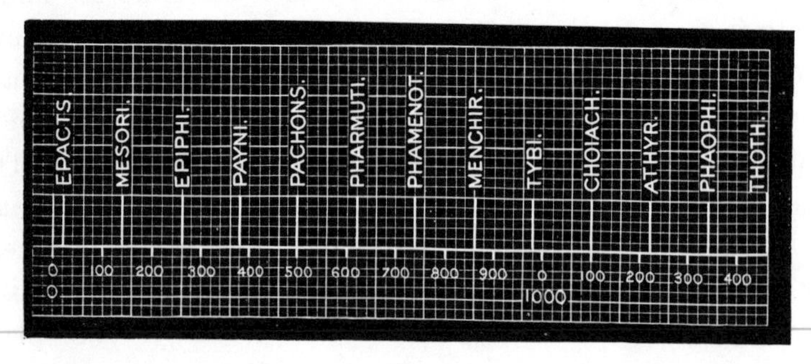

THE DISTRIBUTION OF THE 1ST OF THOTH (REPRESENTING THE RISE OF SIRIUS) AMONG THE EGYPTIAN MONTHS IN THE 1460-YEAR SOTHIC CYCLE.

end of the fixed year, and that is the reason I have reversed the order of months in the diagram.

Now it is clear that, if the Egyptians really worked in this fashion, the date of the heliacal rising of Sirius, given in this way, would enable us to determine the number of years which had elapsed from the beginning of the cycle.

This calendar system, it will be seen, is good only for groups of four years. Now, a system which went no further than this would be a very coarse one. We find, however, that special precautions were taken to define which year of the four was in question, and the fact that this was done goes some way to support the suggestion I have made. Brugsch,[1] indeed, shows that a special sign was employed to mark the first year of each series of four.

Next, as a matter of fact it is known (I have the high authority of Dr. Krall for the statement) that each king was supposed to begin his reign on the 1st Thoth (or 1st Pachons) of the particular year in which that event took place, and the fact that this was so further supports the suggestion we are considering. During the reign its length and the smaller events might be recorded in vague years and days so long as the date of its commencement had been referred to a cycle.

The diagram will show how readily the cycle year can be determined for any vague year. If, for instance, the 1st Thoth in the vague year falls on 1 Tybi of the cycle, we see that 980 years must have elapsed since the beginning of the cycle, and so on.

Here, then, we have a true calendar system. If the Egyptians had not this, what had they?

Dealing, then, with the matter so far as we have gone, we find that the system suggested enabled the place of the

[1] "Matériaux pour servir à la Reconstruction du Calendrier," p. 29.

beginning of each vague year and of each king's reign to be dated in terms of the cycle of 1460 years; and further that, if they had not such a system as this, they had no means of recording any lapse of time which exceeded a year. It is not likely that any nation would put itself in such a position, least of all the ancient Egyptians.

The existence of periods of 365 years and of 120 years among the Egyptians is easily explained when the existence of this great year is recognised; the 365 years' period, marking approximately the intervals from solstice to equinox and equinox to solstice, in the natural year.

Let us next try to get a little further by assuming the supposed method of dating to have been actually employed, and finding the year of the beginning of one or more of these cycles thus obtained. This should eventually help us to determine whether or not the Egyptians acted on this principle, or used one widely different. In such an investigation as this, however, we are terribly hampered by the uncertainty of Egyptian dates; while, as I have said before, there is great divergence of opinion among Egyptologists as to whether, from very early times, there was not a true fixed year.

But let us suppose that the vague year was in common use as a civil year, and that the rising of Sirius started the year; then, if we can get any accepted date to work with, and use the diagram to see how many years had elapsed between that date and the start-point of the cycle, we shall see if there be any cyclical relation; and if we find it, it will be evidence, so far as it goes, of the existence not only of a vague year, but of the mode of reckoning we are discussing.

Now it so happens that there are three references, with dates given, to the rising of Sirius in widely different times; and, curiously enough, the month references are nearly the same. I begin with the most recent, as in this case the date can be fixed with the greater certainty. It is an inscription at Philæ, described by Brugsch (p. 87), who states that, when it was written, the 1st of Thoth = 28th of Epiphi. That is, according to the view we are considering, the heliacal rising of Sirius—that is, the 1st Thoth of the vague year fell on the 28th Epiphi of our cycle. He fixes the date of the inscription between 127 and 117 B.C. Let us take it as 122. Next, referring to our diagram to find how many years had elapsed since the beginning of the cycle, we have—

Days.
5 Epacts.
30 Mesori.
2 Epiphi.

37 × 4 = 148 years elapsed.

The cycle, then, began in (148 + 122 =) 270 B.C.

We next find a much more ancient inscription recording the rising of Sirius on the 28th of Epiphi. Obviously, if the Sothic cycle had anything to do with the matter, this must have happened 1458 years earlier, *i.e.*, about (1458 + 122 =) 1580 B.C. Under which king? Thothmes III., who reigned, according to Lepsius, 1603–1565 B.C.; according to Brugsch, 1625–1577. Now, the inscription in question is stated to have been inscribed by Thothmes III., and, it may be added, on the temple (now destroyed) at Elephantine.

There is yet another inscription, also known to be of a still earlier period, referring to the rising of Sirius on the 27th of Epiphi. We may neglect the difference of one day in the cycle (representing four years); and again, if the use of the Sothic cycle were the origin of the identity of dates, we have this time, according to Oppolzer, a period of 1460 years to add: this gives us (1580 + 1460 =) 3040 B.C. Again under which king? Here we are face to face with one of the difficulties of these inquiries, to which reference has already been made. It may be stated, however, that the inscription is ascribed to Pepi, and that, according to various authorities, that king reigned some time between 3000 and 3700 B.C.

We come, then, to this: that one of the oldest dated inscriptions known seems to belong to a system which continued in use at Philæ up to about 100 B.C., and it was essentially a system of a vague year, the 1st Thoths of which were represented as days on a 1460-years' cycle.

Now, assuming that the approximate date of the earliest inscription is 3044 B.C., and that it represented the heliacal rising of Sirius on the 27th of Epiphi, the year 3044 must have been the [(5 + 30 + 3) × 4 =] 152nd after the beginning of the cycle. The cycle, then, must have commenced (3044 + 152 =) 3196 B.C.

If we assume that the real date of Pepi, who, it is stated, reigned 100 years, included the year 3044 B.C., it may be, then, that the inscriptions to which I have directed attention give us three Sothic cycles beginning—

122 + 148 = 270 B.C.
1580 + 148 = 1728 B.C.
3044 + 148 = 3192 B.C.

According to Biot's calculation, the first heliacal rising of Sirius at the solstice took place in the year 3285 B.C.; it is possible, then, that the Egyptians utilised this heliacal rising within a hundred years of the date on which it would have been first possible for them to do so. This shows how keenly alive they were in these matters, and also, I think, that they had been trained by watching some other star previously.

It would also follow that the vague year was in common

use. There is ample evidence to show, however, that by this time the priests were fully acquainted with the true year, which was called the sacred year, and that every four years an additional epact was interpolated. Their solar temples, then, at last had been utilised.

One argument which has been used to show that a vague year was not in use during the time of the Ramessids has been derived from some inscriptions at Silsilis which refer to the dates on which sacred offerings were presented there to the Nile-god. As the dates 15th of Thoth and 15th of Epiphi are the same in all three inscriptions, although they cover the period from Rameses II. to Rameses III.—120 years —it has been argued by Brugsch that a fixed year is in question.

Brugsch points out that the two dates are separated by 65 days; that this is the exact interval between the Coptic festivals of the commencement of the flow and the marriage of the Nile—the time of highest water; and that, therefore, in all probability these are the two natural phenomena to commemorate which the offerings on the dates in question were made.

But Brugsch does not give the whole of the inscription. A part of it, translated by De Rougé,[1] runs thus :—

> "I (the king) know what is said in the depôt of the writings which are in the House of the Books. The Nile emerges from its fountains to give the fulness of life-necessaries to the gods," etc.

De Rougé justly remarks:

> "Le langage singulier que tient le Pharaon dédicateur pourrait même faire soupçonner *qu'il ne s'agit pas de la venue effective de l'eau sainte du Nil à l'une des deux dates précitées.*"

Krall (*loc. cit.*, p. 51) adds the following interesting remarks :—

[1] "Aeg. Zeit.," 1886, p. 5, quoted by Krall.

"Consider, now, what these 'Scriptures of the House of Life' were like. In a catalogue of books from the temple of Edfû we find, besides a series of purely religious writings, 'The knowledge of the periodical recurrence of the double stars (sun and moon),' and the 'Law of the periodical recurrence of the stars."

" . . . The knowledge embodied in these writings dated from the oldest times of the Egyptian empire, in which the priests placed, rightly or wrongly, the origin of all their sacred rolls" (*cf.* Manetho's "History," p. 130).

Now, to investigate this question we have to approach some considerations which at first sight may seem to be foreign to our subject. I shall be able to show, however, that this is not so.

Imprimis we must remember that it is a question of Silsilis, where we know, both from tradition and geological evidence, in ancient times the first cataract was encountered. The phrase "the Nile emerges from its fountains" would be much more applicable to Silsilis, the seat of a cataract, than as it is at present. We do not know when the river made its way through this impediment, but we do know that after it took place and the Nile stream was cleared as far as the cataract that still remains at Elephantine, a nilometer was erected there, and that during the whole of later Egyptian history, at all events, the time of the rise of the river has been carefully recorded both there and at Rôda.

From this it is fair to infer that in those more ancient times the same thing took place at Silsilis; if this were so, the reason of the record of the coming of the inundation at Silsilis is not far to seek, and hence the suggestion lies on the surface that the records in question may state the date of the arrival *in relation to Memphis time.*

It has been rendered, I hope, quite clear in Chapter XXIII. that there is a difference of fifteen or sixteen days between the arrival of the inundation at Elephantine and at Memphis.

Hence, if in Pepi's time a Nile rise were observed at Silsilis, there might easily be a difference of fifteen days between the rise of the Nile at Silsilis and the Memphic 1st of Thoth. If both at Silsilis and Memphis the Nile rise marked 1st Thoth, the day of the rise at Memphis would correspond to 15th Thoth at Silsilis, so that a king reaching Silsilis with Memphis local time would be struck with this difference, and anxious to record it. May not this, then, have been the important datum recorded in the sacred books? If so, it would not touch the question of the fixed or vague year at all.

Let it, then, be for the present conceded that there was a vague year, and that at least some of the inscriptions which suggest the use of only a fixed year in these early times may be explained in another way.

CHAPTER XXVII.

THE CALENDAR AND ITS REVISION.

In the last chapter the so-called Sothic cycle was discussed, and dates of the commencement of the successive cycles were suggested.

These dates were arrived at by taking the very simplest way of writing a calendar in pre-temple times, and using the calendar inscriptions in the most natural way.

The dates for the coincidence of the heliacal rising of Sirius and the 1st Thoth of the vague year at, or near, the solstice, were—

270 B.C.
1728 B.C.
3192 B.C.

Here, *in limine*, we meet with a difficulty which, if it cannot be explained, evidently proves that the Egyptians did not construct and use their calendar in the way we have supposed.

We have it on the authority of Censorinus that a Sothic period was completed in 139 A.D., and that there was then a vague year in partial use. It is here that the work of Oppolzer is of such high value to us; he discussed all the statements made by Censorinus, and comes to the conclusion that his account is to be depended upon. It has followed from the inquiries of chronologists that in this year the 1st of Thoth took place on July 20 (Julian), the date originally of the heliacal rising of Sirius, the beginning of the year.

This being so, then, in the year 23 A.D.—in which the

Alexandrine reform of the calendar, of which more presently, was introduced—the 1st of Thoth would take place on August 29, a very important date. Censorinus also said that in his own time (A.D. 238) the 1st of Thoth of the vague year fell on

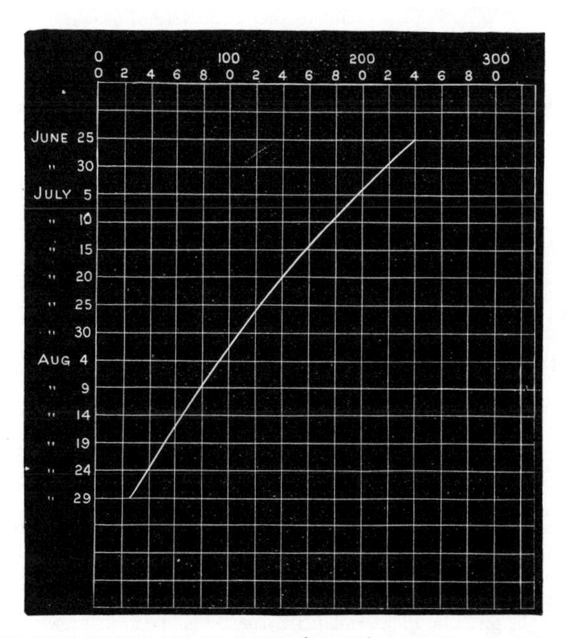

JULIAN DATES OF THE 1ST OF THOTH (VAGUE) FROM 23 A.D. AND 240 A.D.

June 25. The diagram will show the connection of these three dates in reference to the vague year. The relations of the statements made as to the years 139 and 238 are very clearly discussed by Prof. Oppolzer.

Oppolzer, then, being satisfied as to the justice of taking the year 139 A.D. as a time of coincidence of the fixed and vague years—the latter being determined alone by the heliacal rising of Sirius, and, be it remembered, not by the solstices—calculated with great fulness, using Leverrier's modern values, the years in which, in the various Egyptian latitudes, chiefly taking Memphis (lat. 30°) and Thebes (lat. 25°), the coincidence between the two Thoths occurred in the previous periods of Egyptian history. He finds these dates for latitude 30° as follow:—

		Julian year.		Historical year.[1]
0	...	−4235	...	−4236
1	...	−2774	...	−2775
2	...	−1316	...	−1317
3	...	+139	...	+139
4	...	+1591	...	+1591
5	...	+3039	...	+3039

Now, the date which Oppolzer gives for the coincidence which is nearest the date we had previously determined at 270 B.C. is 139 A.D. There is a difference of 409 years.

The question is, Can this fundamental difference be explained? I think it can.

[1] It should be observed that a distinction is made between the Julian and the historical year. This comes from the fact that when astronomical phenomena are calculated for dates B.C., it must be remembered that chronologists are in the habit of designating by 1, or rather by —1, the first year which precedes the instant of time at which the chronological year commenced, while astronomers mark this year in their tables by 0. It follows, therefore, that the rank of any year B.C. is always marked by an additional unit in the chronological dates. For the Christian era, of course, chronologists and astronomers work in the same way. The following table, given by Biot, exhibits the connection between these two methods. In the latter Biot shows the leap-years marked B, and the corresponding years in the Scaligerian chronological period are also given.

DATES OF JULIAN YEARS COMMENCING ON JANUARY 1.

According to Chronologists.		According to Astronomers.		Corresponding years of the period of Scaliger.
—6	... ...	—5	... ...	4708
—5B	... ...	—4B	... ...	4709
—4	... ...	—3	... ...	4710
—3	... ...	—2	... ...	4711
—2	... ...	—1	... ...	4712
—1B	... ...	—0B	... ...	4713B
Physical instant when the era commenced.				
+1	... ...	+1	... ...	4714
+2	... ...	+2	... ...	4715
+3	... ...	+3	... ...	4716
+4B	... ...	+4B	... ...	4717B
+5		+5		4718

In the first place, it is beyond doubt that, in the interval between the Ramessids and the Ptolemies, the calendar, even supposing the vague year to have been used and to have been retained, had been fundamentally altered, and the meanings of the hieroglyphics of the tetramenes had been changed—in other words, the designations of the three seasons had been changed.

On this point I quote Krall:—

"It is well known that the interpretation of the seasons and the months given by Champollion was opposed by Brugsch, who propounded another, which is now universally adopted by experts. Something has happened here which is often repeated in the course of Egyptian history—the signs have changed their meaning. Under the circumstance that the vague year during 1461 years wanders through the seasons in a great cycle, it is natural that the signs for the tetramenes should have changed their significations in the course of millenniums.

"While Thoth was the first month of the inundation in the documents of the Thutmosids and Ramessids, we have in the time of the Ptolemies the month Pachons as the first month of the flood season. Whilst Brugsch's explanation is valid for the time of the Ramessids, it is not so for that of the Ptolemies, to which Champollion's view is applicable."

The signs used for the tetramenes are supposed to represent water, a field with growing plants, and a barn; the natural order would be that the first should represent the inundation, the second the sowing which succeeds it, and the last harvest-time. If this be conceded, the initial system would have had the month Thoth connected with the water sign, as Thoth in early Egyptian times was the first inundation month. But in the times of the Ramessids even this is not so. Thoth has the sowing sign assigned to it. In the time of the Ptolemies the flood is no longer in Thoth, but in Pachons, and Pachons has the barn sign attached to it, while the month

[1] *Loc. cit.*, p. 29.

Thoth is marked by the water sign, thereby bringing back the hypothetical relation *between the name of the month and the sign*, although, as we have seen, Thoth is no longer the flood month.

Egyptologists declare that all, or at least part, of this change took place between the periods named; they are undoubtedly justified as regards a part.

At one point in this interval we are fortunately supplied with some precise information. In the year 238 B.C. a famous decree was published, variously called the decree of Canopus and the decree of Tanis, since it was inscribed on a stone found there. It is perfectly clear that one of the functions of this decree was to change, or to approve an already made change in, the designation of the season or tetramene in which the inundation commenced, from Thoth to Pachons.

Another function was to establish a fixed year, as we shall see presently. We must assume, then, that a vague year was in vogue prior to the decree. Now the decree tells us that at its date the heliacal rising of Sirius took place on 1 Payni. Assuming that this date had any relation to the system we have been considering, the cycle to which it belonged must have begun

Days.	
5	Epacts
30	Mesori
30	Epiphi
30	Payni
—	
95 × 4 = 380 years previously—that is, in the year 618 B.C.	

Here at first sight it would seem that the Sothic cycles we have been considering have no relation to the one now in question; for, according to my view, the last Sothic cycle began in 1728 B.C. A little consideration, however, will lead to the contrary view, and show that the time about 600 B.C. was very convenient for a revision of the calendar.

In the first place, nearly a month now elapsed between the coming of the flood and the heliacal rising; and in the second, by making the year for the future *to begin with the flood*, a change might be made involving tetramenes only.

Thus, commencement of cycle ...		1728 B.C.
Epacts	5	
Two tetramenes ...	240	
Month between flood and rising of Sirius ...	30 [1]	
	275 × 4 =	1100
		628 B.C.

Nor is this all. A very simple diagrammatic statement will

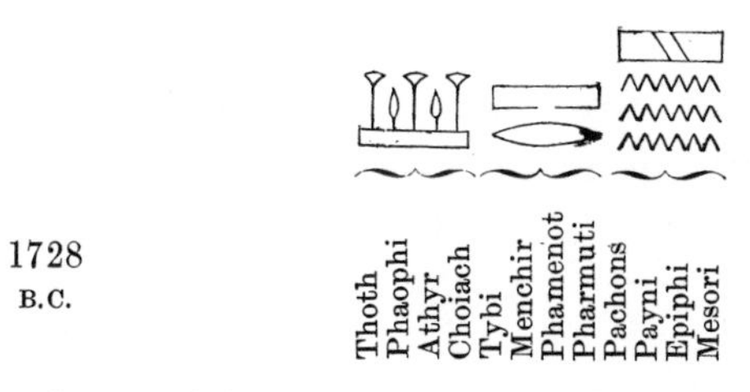

show what might also have happened about 618 B.C. if a reformer of the calendar (and one especially of conservative tendencies) appeared upon the scene, who believed that the ancient sign for the inundation-tetramene was the water sign, and that the ancient name was Thoth. Finding the cycle beginning in 1728 B.C. with the signs as shown above—

B.C.
618

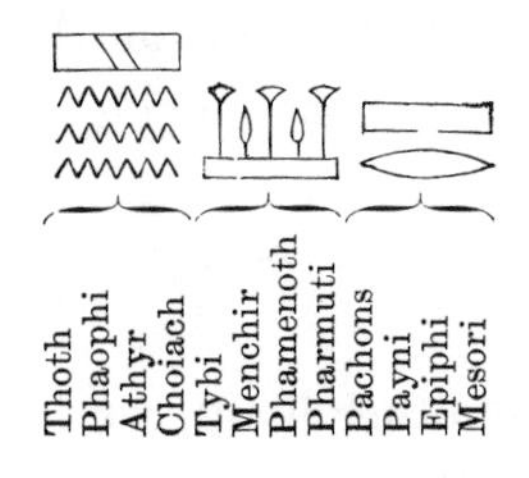

[1] Probably too great a value by two or three days.

when starting fresh, he would seize the opportunity of effecting a change, not only by dealing with a tetramene, but he would change the names of the tetramenes allocated to the signs; as Krall remarks, it was almost merely a question of a change of the sign! It really was more, because the new tetramene began with the flood.

Assuming this, we can see exactly what was done in 238 B.C., *i.e.*, about 380 years later. We have seen that the 380 years is made up of

5 Epacts
30 Mesori
30 Epiphi
30 Payni
—
95 × 4 = 380

—the heliacal rising of Sirius occurring on 1 Payni, having swept backwards along the months in the manner already explained. We had, to continue the diagrammatic treatment—

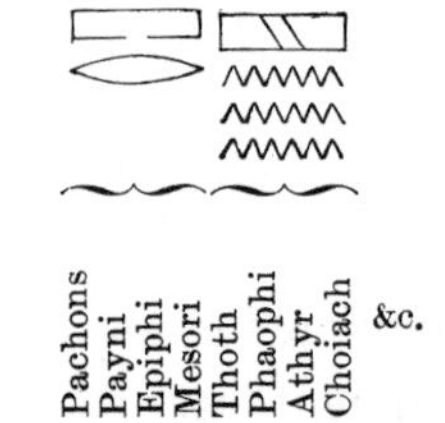

B.C.
238

To sum up, so far as we have gone, we have the three inscriptions at Philæ, Elephantine and the still more ancient one of Pepi (?), indicating on the simple system we have suggested beginnings of Sothic cycles on the 1st Thoth about the years

270
1728 } B.C.
3192

On the other hand, we have the decree of Canopus, giving us by exactly the same system a local revision of the calendar about 600 B.C. I say *about* 600 B.C. because it must be remembered that a difference of $2\frac{1}{2}$ days in the phenomena observed will make a difference of 10 years in the date, and we do not know in what part of the valley the revision took place, and therefore at what precise time in relation to the heliacal rising the Nile-rise was observed.

Whenever presumably it took place, New Year's Day was reckoned by the Flood, and the rising of Sirius followed nearly, if not quite, a month afterwards. The equivalent of the old 1st Thoth was therefore 1 Payni. In months, then, the old 1st Thoth was separated from the new one (= 1 Payni) by three months (Payni, Mesori, Epiphi) and the Epacts.

In this way, we can exactly account for the difference of 409 years referred to above as the dates assigned by Censorinus and myself for the beginning of the Sirius cycle.

Difference between 270 and 239	=	31 years.
3 months = 90 days × 4	=	360 ,,
5 epacts × 4	=	20 ,,
		411 ,,

The difference of two years is equal only to half a day!

It seems, then, pretty clear from this that the suggestion I have ventured to make on astronomical grounds may be worth consideration on the part of Egyptologists. If our inquiries have really led us to the true beginnings of the Sothic periods, it is obvious that those who informed Censorinus that the year 139 A.D. was the end of a cycle *omitted to tell him what we now can learn from the decree of Tanis.*

CHAPTER XXVIII.

THE FIXED YEAR AND FESTIVAL CALENDARS.

THE reformation of the Egyptian calendar, to be gathered, as I suggested in the last chapter, from the decree of Tanis, is not, however, the point to which reference is generally made in connection with the decree. The attempt recorded by it to get rid of the vague year is generally dwelt on.

Although the system of reckoning which was based on the vague year had advantages with which it has not been sufficiently credited, undoubtedly it had its drawbacks.

The tetramenes, with their special symbolism of flood-, seed-, and harvest-time, had apparently all meant each in turn; however the meanings of the signs were changed, the "winter season" occurred in this way in the height of summer, the "sowing-time" when the whole land was inundated and there was no land to plant, and so on. Each festival, too, swept through the year. Still, it is quite certain that information was given by the priests each year in advance, so that agriculture did not suffer; for if this had not been done, the system, instead of dying hard, as it did, would have been abolished thousands of years before.

Before I proceed to state shortly what happened with regard to the fixing of the year, it will be convenient here to state a suggestion that has occurred to me, on astronomical grounds, with regard to the initial change of sign.

It is to be noted that in the old tables of the months, instead of Sirius leading the year, we have Teχi with the two feathers of Amen. In later times this is changed to Sirius.

I believe it is generally acknowledged that the month-table at the Ramesseum is the oldest one we have; there is a variant at Edfû. They both run as follows, and no doubt they had their origin when a 1st Thoth coincided with an heliacal rising and Nile flood.

Egyptian month.	Tropical month.	Ramesseum.	Edfû.
1. Thoth	June—July	Teχi	Teχ
2. Phaophi	July—Aug.	Ptah (Ptah-res-aneb-f)	Ptah Menχ
3. Athyr	Aug.—Sep.	Hathor	?
4. Choiach	Sep.—Oct.	Paχt	Kehek
5. Tybi	Oct.—Nov.	Min	Set-but
6. Menchir	Nov.—Dec.	Jackal (rekh-ur)	Hippopotamus (rekh-ur)
7. Phamenoth	Dec.—Jan.	,, (rekh-netches)	Hippopotamus (rekh-netches)
8. Pharmuthi	Jan.—Feb.	Rennuti	Renen
9. Pachons	Feb.—Mar.	χensu	χensu
10. Payni	Mar.—Ap.	Horus (χonti)	Horus (Hor-χent-χati)
11. Epiphi	Ap.—May	Åpet	Apet
12. Mesori	May—June	Horus (Hor-m-χut)	Horus (Hor-ra-m-χut)

I am informed that Texi, in the above month-list, has some relation to Thoth. In the early month-list the goddess is represented with the two feathers of Amen, and in this early stage I fancy we can recognise her as Amen-t; but in later copies of the table the symbol is changed to that of *Sirius*. This, then, looks like a change of cult depending upon the introduction of a new star—that is, a star indicating by its heliacal rising the Nile-rise after the one first used had become useless for such a purpose.

I have said that the Ramesseum month-list is probably the oldest one we have. It is considered by some to date only from Rameses II., and to indicate a fixed year; such, however, is not Krall's opinion.[1] He writes:—

[1] *Op. cit.*, p. 48.

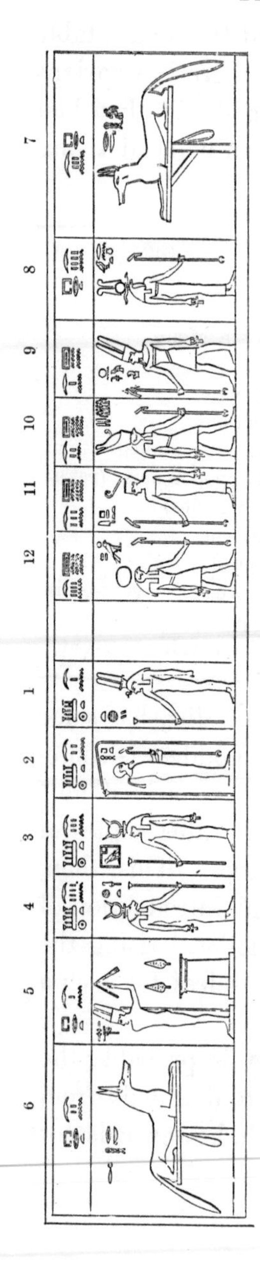

THE MONTH-TABLE AT THE RAMESSEUM.

"The latest investigations of Dümichen show that the calendar of Medînet-Habû is only a copy of the original composed under Ramses II. about 120 years before. . . . .

"But the true original of the calendar of Medînet-Habû does not even date from the time of Ramses II. It is known to every Egyptologist how little the time of the Ramessids produced what was truly original, how much just this time restricted itself to a reproduction of the traditions of previous generations. In the calendar of Medînet-Habû we have (p. 48) not a fixed year instituted under Ramses II., but the normal year of the old time, the vague year, as it was, to use Dschewhari's words quoted above (p. 852) in the first year of its institution, the year as it was before the Egyptians had made two unwelcome observations: First, that the year of 365 days did not correspond to the reality, but shifted by one day in four years with regard to the seasons; secondly—which, of course, took a much longer time—that the rising of Sirius ceased to coincide with the beginning of the Nile flood.

"We are led to the same conclusion by a consideration of the festivals given in the calendar of Medînet-Habû. They are almost without exception the festivals which we have found in our previous investigation of the calendars of Esne and Edfû to be attached to the same days. We know already the Uaya festival of the 17th and 18th Thoth, the festival of Hermes of the 19th Thoth, the great feast of Amen beginning on the 19th Paophi, the Osiris festivals of the last decade of Choiak, and that of the coronation of Horuz on the 1st Tybi.

"Festivals somehow differing from the ancient traditions and general usage are unknown in the calendar of Medînet-Habû, and it is just such festivals which have enabled us to trace fixed years in the calendars of Edfû and Esne.

"We are as little justified in considering the mythologico-astronomical representations and inscriptions on the graves of the time of the Ramessids as founded on a fixed year, as we can do this in the case of the Medînet-Habû calendar. In this the astronomical element of the calendar is quite overgrown by the mythological. Not only was the daily and yearly course of the sun a most important event for the Egyptian astronomer, but the priest also had in his sacred books many mythological records concerning the god Rā, which had to be taken into account in these representations. The mythological ideas dated from the oldest periods of Egyptian history; we shall therefore be obliged, for their explanation, not to remain in the thirteenth or fourteenth century before Christ, but to ascend into previous centuries. *I should think about the middle of the fourth millennium before Christ, that is the time at which the true original of the Medînet-Habû calendar was framed.* Further, we must in these mythological and astronomical representations not overlook the fact that we cannot expect them to show mathematical accuracy—that, on the contrary, if that is a consideration, we must proceed with the greatest caution. We know now how inexact were the representations and texts of tombs, especially where the Egyptian artist could suppose that no human eye would inspect his work; we also know how often representations stop short for want of room, and how much the contents were mutilated for the sake of symmetry."

There is also, as I have indicated, temple evidence that Sirius was not the first star utilised as a herald of sunrise. We have, then, this possibility to explain the variation from the

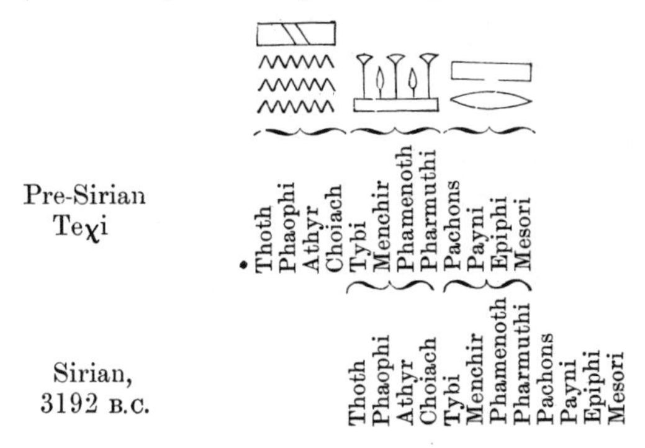

SYRIAN AND PRE-SYRIAN TETRAMENE-SIGNS.

true meaning of the signs in Ramessid times. And it may be gathered from this that the calendar was reorganised[1] when

[1] Goodwin has already asked, "Does the Smith Papyrus refer to some rectification of the calendar made in the fourth dynasty, similar to that made in Europe from the old to the new style?" Quoted by Riel, "Sonnen- und Sirius-Jahr," p. 361.

the Sirius worship came in, and that the change effected in 619 B.C. brought the hieroglyphic signs back to their natural meaning and first use.

The whole story of calendar revision may, therefore, possibly have been as follows :—

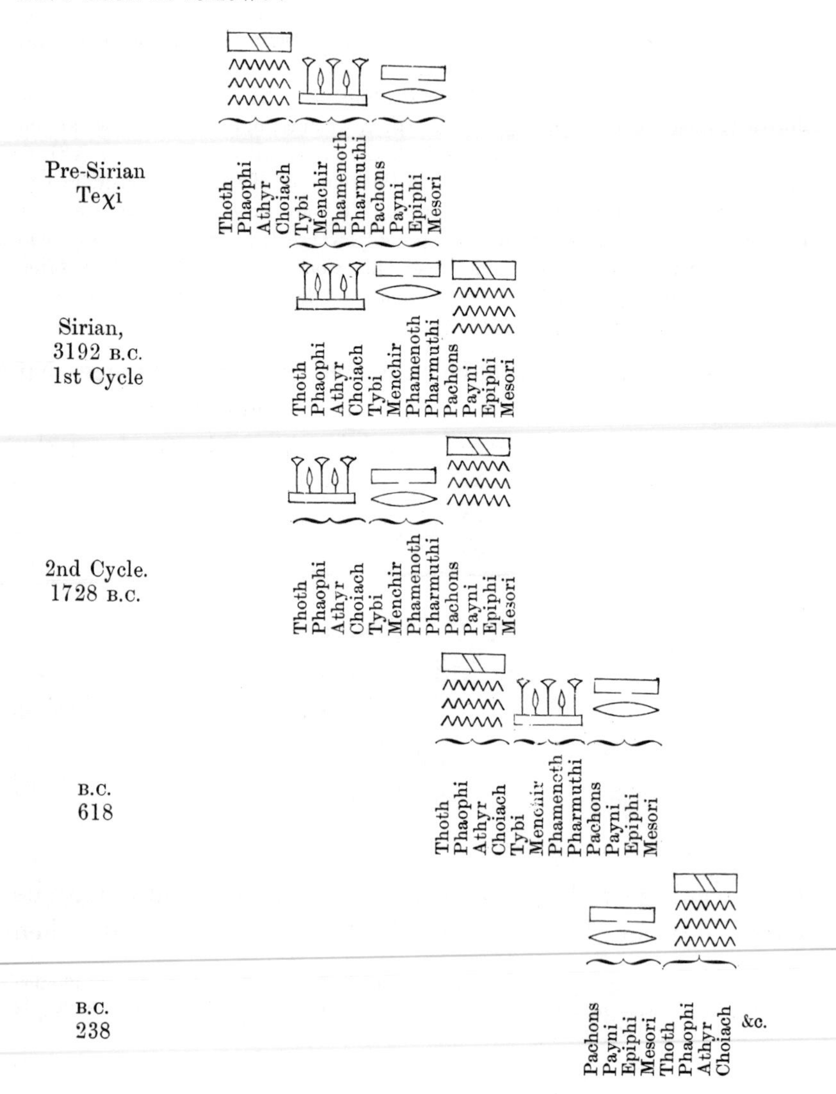

The revision of 618 B.C. was not universally accepted, so from that time onward there was an old and a new style in force.

Before I pass on, it may be convenient, in connection with the above month-tables, to refer in the briefest way to the mythology relating to the yearly movement of the sun, in order to show that when this question is considered at all, if it helps us with regard to the mythology connected with the rising and setting of stars, it will as assuredly help us with regard to the mythology of the various changes which occur throughout the year.

We have, as we have seen, in the Egyptian year really the prototype of our own. The Egyptians, thousands of years ago, had an almost perfect year containing twelve months; but, instead of four seasons, they had three—the time of the sowing, the time of the harvest, and the time of the inundation. Unfortunately, at various times in Egyptian history, the symbols for the tetramenes seem to have got changed.

The above-given inscriptions show that they had a distinct symbolism for each of the months. Gods or goddesses are given for ten months out of the twelve, and where we have not these we have the hippopotamus (or the pig) and the jackal, two circumpolar constellations. I think there is no question that we are dealing here with these constellations, though the figures have been supposed to represent something quite different.

There are also myths and symbols of the twelve changes during the twelve hours of the day; the sun being figured as a child at rising, as an old man when setting in the evening. These ideas were also transferred to the annual motion of the sun. In Macrobius, as quoted by Krall, we find the statement that the Egyptians compared the yearly course of the sun also with the phases of human life.

Little child	=	Winter Solstice.
Young man	=	Spring Equinox.
Bearded man	=	Summer Solstice.
Old man	=	Autumnal Equinox.

With the day of the Summer Solstice the sun reaches the greatest northern rising amplitude, and at the Winter Solstice its greatest southern amplitude. By the solstices the year is divided into two approximately equal parts; during one the points of rising move southwards, during the other northwards.

This phenomenon, it is stated, was symbolised by the two eyes of Rā, the so-called Utchats, which look in different directions. They appear as representing the sun in the two halves of the year.

We have next to discuss the fixed year, to which the Egyptian chronologists were finally driven in later Egyptian times. The decree of Tanis was the true precursor of the Julian correction of the calendar. In consequence of this correction *we* now add a day every four years to the end of February. The decree regulated the addition, by the Egyptians, of a day every four years by adding a day to the epacts, which were thus six every four years instead of being always five, as they had been before.

In fact, it replaced the vague year by the sacred year long known to the priests.

But if everything had gone on then as the priests of Tanis imagined, the Egyptian New Year's Day, *if* determined by the heliacal rising of Sirius, would not always afterwards have been the 1st of Payni, although the solstice and Nile flood would have been due at Memphis about the 1st of Pachons; and this is, perhaps, one among the reasons why the decree was to a large extent ignored.

Hence, for some years after the date of the decree of Tanis, there were at least three years in force—the new fixed year, the new vague year, reckoning from Pachons, and the old vague year, reckoning from Thoth.

But after some years another attempt was made to get rid of all this confusion. The time was 23 B.C., 216 years after the decree of Tanis, and the place was Alexandria. Hence the new fixed year introduced is termed the Alexandrine year.

This new attempt obviously implied that the first one had failed; and the fact that the vague year was continued in the interval is sufficiently demonstrated by the fact that the new year was $\frac{216}{4} = 54$ days *en retard.* In the year of Tanis it is stated that the 1st Pachons, the new New Year's Day, the real beginning of the flood, fell on the 19th of June (Gregorian), the Summer Solstice, and hence the 1st of Thoth fell on the 22nd of October (Gregorian). In the Alexandrine year the 22nd of October is represented by the 29th of August, and the 19th of June by the 20th of April.

It is noteworthy that in the Alexandrine year the heliacal rising of Sirius on the 23rd of July (Julian) falls on the 29th of Epiphi, nearly the same date as that to which I first drew attention in the inscriptions of the date of Thothmes and Pepi. This, however, it is now clearly seen, is a pure accident, due to the break of continuity before the Tanis year, and the *slip* between that and the Alexandrine one. It is important to mention this, because it has been thought that somehow the "Alexandrine year" was in use in Pepi's time.

It would seem that the Alexandrine revision was final,

and that the year was truly fixed, and from that time to this it has remained so, and must in the future for ever remain so. It must never be forgotten that we owe this per fection to the Egyptian Festival Calendars.

One of the chief uses of the Egyptian calendar that has come down to us was the arrangement and dating of the chief feasts throughout the year in the different temples.

The fact that the two great complete feast-calendars of Edfû and Esne refer to the only fixed years evidenced by records—those of Tanis and Alexandria—one of which was established over 200 years after the other, is of inestimable value for the investigation of the calendar and chronology of ancient Egypt.

In an excellent work of Brugsch, "Three Festival Calendars from the Temple of Apollinopolis Magna (Edfû) in Upper Egypt," we have two calendars which we can refer to fixed years, and can date with the greatest accuracy. In the case of one of these, that of Esne, this is universally recognised; as to the other, that of Apollinopolis Magna, we are indebted to the researches of Krall, who points out, however, that "it is only when the province of Egyptian mythology has been dealt with in all directions that we can undertake a successful explanation of the festival catalogues. Even externally they show the greatest eccentricities, which are not diminished, but increased, on a closer investigation."

About some points, however, there is no question. The Summer Solstice is attached in the Edfû calendar to the 6th Pachons, according to Krall, while the beginning of the flood is noted on the 1st of that month. In the Esne calendar the 26th Payni is New Year's Day. We read:—"26th Payni, New Year's Day, Feast of the Revelation of Kahi in the Temple. To dress the crocodiles, as in the month of Menchir, day 8."

Peculiar to the Esne calendar, according to Krall, is the mentioning of the "New Year's Festival of the Ancestors" on the 9th of Thoth; to the Edfû calendar, publication No. 1 of Brugsch, the festival "of the offering of the first of the harvested fruits, after the precept of King Amenemha I.," on the 1st Epiphi, and "the celebration of the feast of the Great Conflagration" on the 9th of Menchir. In feast-calendar No. 1, the reference to the peculiar Feast of Set is also remarkable; this was celebrated twice, first in the first days of Thoth (? 9th), then, as it appears, in Pachons (10th). This feast is well known to have been first mentioned under the old Pharaoh Pepi Merinrā.

It is a question whether in the new year of the ancestors and the feasts of Set, all occurring about the 9th Thoth and Pachons, we have not Memphis festivals which gave way to Theban ones; for, so far as I can make out, the flood takes about nine days to pass from Thebes to Memphis, so that in Theban time the arrival of the flood at Memphis would occur on 9th or 10th Thoth. There is no difficulty about the second dating in Pachons, for, as we have seen, this followed on the reconstruction of the calendar.

It is also worthy of note that the feast of the "Great Conflagration" took place very near the Spring Equinox.

Let us dwell for a moment on the Edfû inscriptions to see if we can learn from them whether or not they bear out the views brought forward with regard to this reconstruction.

As we have seen, it is now acknowledged that the temple inscriptions at Edfû (which are stated to have been cut between 117 and 81 B.C.[1]) are based upon the fixed year of Tanis; hence we should expect that the rising of Sirius would be referred to

[1] On the 7th Epiphi of the tenth year of Ptolemy III. the ceremony of the stretching off the cord took place. Dümichen, *Aeg. Zeit.*, 2, 1872, p. 41.

on 1 Payni, and this is so. But here, as in the other temples, we get double dates referring to the old calendars, and we find the "wounding of Set" referred to on the 1st Epiphi and the rising of Sirius referred to under 1 Mesori. Now this means, if the old vague year is referred to, as it most probably is, that

$$\begin{array}{l} 5 \text{ Epacts} \\ 30 \text{ Mesori} \\ \hline 35 \times 4 = 140 \text{ years} \end{array}$$

had elapsed since the beginning of a Sothic cycle, when the calendar coincidences were determined, which were afterwards inscribed on the temple walls. We have, then, 140 years to subtract from the beginning of the cycle in 270 B.C. This gives us 130 B.C., and it will be seen that this agrees as closely as can be expected with my view, whereas the inscription has no meaning at all if we take the date given by Censorinus.

I quote from Krall[1] another inscription common to Edfû and Esne, which seems to have astronomical significance.

"1. Phamenoth. Festival of the suspension of the sky by Ptah, by the side of the god Harschaf, the master of Heracleopolis Magna (A1). Festival of Ptah. Feast of the suspension of the sky (Es).

"Under the 1st Phamenoth, Plutarch, *de Iside et Osiride*, c. 43, b, notices the ἔμβασις Ὀσίριδος εἰς τὴν σελήνην. These are festivals connected with the celebration of the Winter Solstice, and the filling of the Uza-eye on the 30th Menchir. Perhaps the old year, which the Egyptians introduced into the Nile valley at the time of their immigration, and which had only 360 days, commenced with the Winter Solstice. Thus we should have in the 'festival of the suspension of the sky,' by the ancient god Ptah—venerated as creator of the world—a remnant of the time when the Winter Solstice . . . . . marked the beginning of the year, and also the creation."

The reconstruction of the calendar naturally enhanced the importance of the month Pachons; this comes out very clearly from the inscriptions translated by Brugsch. On this point Krall remarks:—

[1] *Op. cit.*, p. 37.

"It is, therefore, quite right that the month Pachons, *which took the place of the old Thoth by the decree of Tanis*, should play a prominent part in the feast-calendars of the days of the Ptolemies, and the first period of the Empire in general, but especially in the *Edfû* calendar, which refers to the *Tanitic* year. The first five days of Pachons are dedicated in our calendar to the celebration of the subjection of the enemies by Horus; we at once remember the above-mentioned (p. 7) record of Edfû of the nature of a mythological calendar, describing the advent of the Nile flood. On the 6th of Pachons—remember the great importance of the sixes in the Ptolemæan records—the solstice is then celebrated. The Uza-eye is then filled, a mythical act which we have in another place referred to the celebration of the solstice, and 'everything is performed which is ordained' in the book 'on the Divine Birth.'"

Next let us turn to Esne. The inscriptions here are stated to be based on the Alexandrine year, but we not only find 1st Thoth given as New Year's Day, but 26 Payni given as the beginning of the Nile flood.

Now I have already stated that the Alexandrine year was practically a fixing of the vague Tanis year—that is, a year beginning on 1st Pachons in 239 B.C.

If we assume the date of the calendar coincidences recorded at Esne to have been 15 B.C. (we know it was after 23 B.C. and at the end of the Roman dominion), we have as before, seeing that, if the vague Tanis year had really continued, it would have swept forward with regard to the Nile flood,

Pachons	30
Payni	26
	—
	56 × 4 = 224 years after 239 B.C.

This double dating, then, proves the continuation of the vague year of Tanis if the date 15 B.C. of the inscription is about right.

Can we go further and find a trace of the old cycle

beginning 270 B.C.? In this case we should have the rising of Sirius

$$\begin{array}{r} 270 \\ -\ 15 \\ \hline 4)255 \text{ years} \\ \hline 64 = \text{say, five Epacts and two months.} \end{array}$$

This would give us 1 Epiphi. Is this mentioned in the Esne calendar? Yes, it is, "1 Epiphi. To perform the precepts of the book on the second divine birth of the child Kahi."

Now the 26th Payni, the new New Year's Day, is associated with the "revelation of Kahi," so it is not impossible that "the second divine birth" may have some dim reference to the feast.

It is not necessary to pursue this intricate subject further in this place; so intricate is it that, although the suggestions I have ventured to make on astronomical grounds seem consistent with the available facts, they are suggestions only, and a long labour on the part of Egyptologists will be needed before we can be said to be on firm ground.

CHAPTER XXIX.

THE MYTHOLOGY OF ISIS AND OSIRIS.

A LONG parenthesis has been necessary in order to inquire fully into the yearly festivals of the Egyptian priests, the relation of the feasts to the rising of stars, and the difficulties which arose from the fact that a true year was not in use till quite late.

It is now time to return to the subject-matter of Chapters XIX.—XXII. in order to show that since the goddesses chiefly worshipped at Denderah and Thebes were goddesses whose cult was associated with the year, it is open to us to inquire whether we may not use the facts with which we are now familiar to obtain a general idea of that part of mythology which refers to them.

I will begin by taking a certain group of goddesses.

1. *There is evidence that many of the goddesses under discussion personified stars in exactly the same way that Isis personified Sirius and Mut γ Draconis.*—If we leave Denderah and Thebes for the moment, and consider the pyramid region of Gîzeh, we find that the temples there, which are associated with each of the pyramids, are not oriented to Sirius; but yet they are temples of Isis, pointing due east; therefore they could not have pointed to the same Isis worshipped at Denderah, or the same Hathor worshipped at Thebes.

Thus, in the case of the temple of Mut at Thebes, of Isis at Denderah, and the temples of Isis at the pyramids, and in many towns facing East, obviously different stars were in question, whatever the mythology might have been.

BLACK GRANITE STATUE OF SEKHET FROM THE TEMPLE OF MUT AT THEBES.

Further, it seems quite certain that the star symbolised as Isis in the pyramid worship was the star Antares (Serk-t) heralding the autumnal equinox, and it is probable that the Pleiades (Nit) were so used at the vernal equinox.

2. *There is evidence that many of the names of these goddesses are pure synonyms.*—That is to say, we have the same goddess (or the same star) called different names in different places, and associated with different animal emblems, in consequence of the existence of

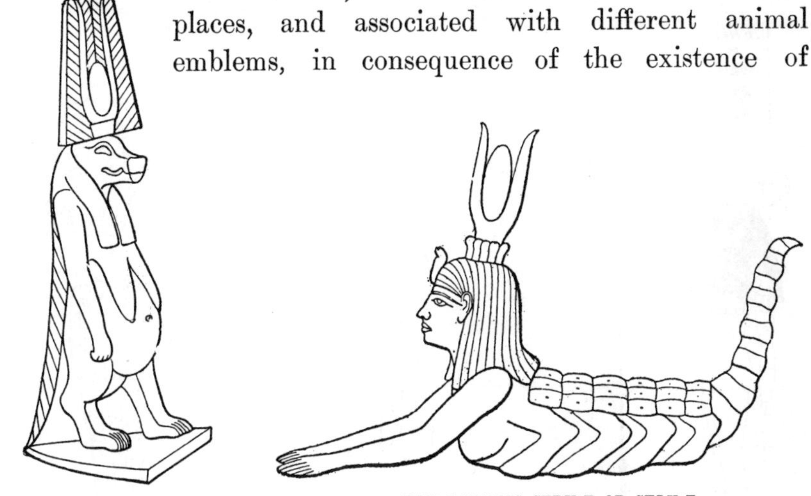

THE GODDESS TAURT. THE GODDESS SERK-T OR SELK-T.
(*Both with horns and disk.*)

different totems in different nomes. I have already referred to the symbolism of the goddess Mut. In one form she is a hippopotamus; in another she has a cow's horns and disk. The temple of Hathor at Denderah was probably associated with the crocodile or the hippopotamus; so that from the symbolism referred to we get the suggestion that the goddess Mut was really the Theban form of the goddess Hathor at Denderah. There is another delineation which shows that even more clearly: it is a drawing of the goddess with both the lion's and crocodile's head. One of the most wonderful things to be seen at Thebes is that marvellous collection of the statues of Sechet in the temple of Mut, all of them lion-headed. From evidence of this kind in addition to the temple inscriptions already referred to, we get a clear indication of the fact that Apet, Mut, Taurt, Sechet, Bast, were the same goddess under different names, and I may add that they, in all probability, symbolised the star γ Draconis.

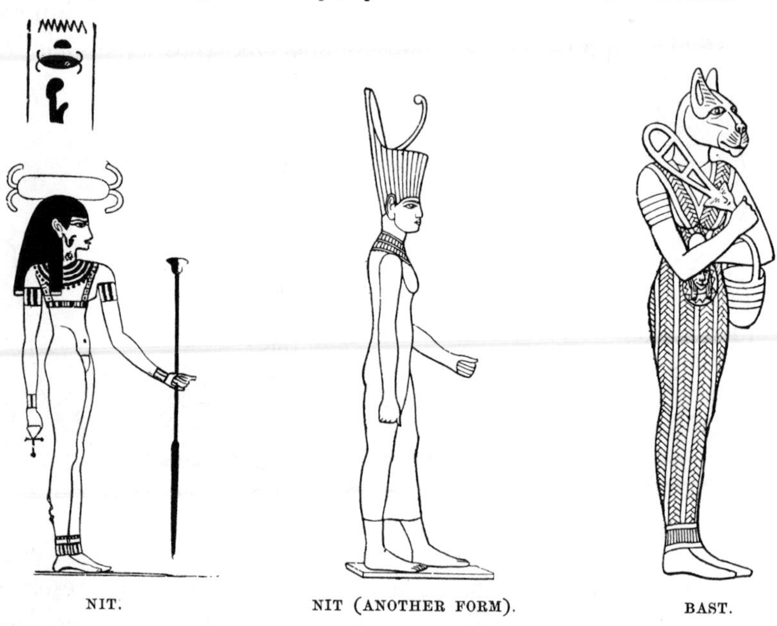

NIT. NIT (ANOTHER FORM). BAST.

3. *All these goddesses have a special symbol.*—Hathor wears the cow's head and the horns with the disk. Taurt, the hippopotamus-goddess, is also represented with horns and disk. The horns and disk are also worn by Serk-t, Sati and Rā-t, the wife of the sun-god Rā; many other goddesses might be added to this list. Indeed, it looks as if all the goddesses who are stated to be variants either of Isis or Hathor have this same symbol.

This generic symbolism suggests that the names Isis and Hathor are themselves generalisations, meaning an accompaniment of sunrise, whether that light be the dawn, or an heliacally-rising star, or even the moon. The generic symbol is the sun's disk and horns, which, I think, may not impossibly be a poetic development of the sign for sunrise. Isis and Hathor are two different ways of defining or thinking about

ANUQA. SATI.

a rising star—that is, a star heralding the sunrise, for such were the rising stars *par excellence.*

All the goddesses so symbolised are either different forms of Isis or Hathor, or represent goddesses who personify or bring before us mythologically stars the rising of which was observed at the dawn at some time of the year or another.

But it must be added that these goddesses are not always

represented with this head-gear, possibly because they had other functions besides their astronomical one.

The extent of this variation may be gathered from the two forms of Neith or Nit given on page 290.

4. *Many of the goddesses are represented as Isis nursing Horus.*—It is very important not to forget that stars were chiefly observed rising in the dawn, and that mythologically such an event was represented by the Egyptians as Isis (the rising star-goddess) nursing Horus (the rising sun-god). The sun

ISIS NURSING HORUS. (The last form is Serk-t-Isis, the scorpion goddess.)

was supposed to be a youth in the morning, to be very young therefore at the moment of rising, and the goddess Isis was supposed to be then nursing him. Many of the goddesses are thus portrayed. I may mention Renen-t, Serk-t, Rā-t, Amen-t, as instances. Thus I hold that we get in this series of goddesses the statement, put mythologically, that certain stars to which the goddesses were sacred rose heliacally at some time of the year or another. Of course the record is far from complete, and probably it will become more complete when inquiries are made from this point of view. The original symbolism is that Isis or Hathor is a star rising in the dawn, watching over the sun or taking him from his cradle; and the young Horus, the rising sun, is, of course, the son of Isis. The emblem of the mother and child is thus shown to have been in established use for the expression of high religious thought at least 5000 years ago.

These and other facts may be brought together in a tabular form, to show what apparently the complete mythology of Isis meant.

ISIS = ANYTHING LUMINOUS TO THE EASTWARD HERALDING SUNRISE.

DAWN.	MOON.	γ DRACONIS. (3000 B.C.)	ANTARES. (3700 B.C.)	α COLUMBÆ. (Before 3000 B.C.)	SIRIUS. (After 3000 B.C.)	DOUBTFUL. (Probably late.)
Isis	Isis	Isis	Serk-t	Teχi	Isis	Anụqa
		Hathor (hawk and hippopotamus)	(N. Egypt)	Amen-t	Hathor (cow)	Hak
					Rā-t	Haka
		Mut (vulture)				Hak-t
		Sechet } Lion or	α Centauri			Hequet
		Bast } cat	3700 B.C.			Maloul
		Menkh	(S. Egypt)			
		Tafnet				
		Apet				
		Nebun				

It will be seen that in the case of Isis we are not dealing merely with a rising star, while, so far as I know, Hathor is limited to stars.

If we accept the general statement regarding Isis, namely, that it was a term applied to anything appearing to the eastward and heralding sunrise, many of our difficulties at once disappear. The Isis of the pyramid-temples and of the smaller temple of Denderah symbolised different celestial bodies, though they served the same purpose. The Hathor of the greater temple of Denderah, and the Hathor of Dêr el-Bahari, symbolised different celestial bodies, but their function was the same. On the other hand, the Hathor of Denderah and the Mut of Thebes were neither different divinities, nor did they personify different stars; they were simply local names of γ Draconis.

We are thus enabled to understand the doubling of the symbolism in the case of Hathor. The hippopotamus and the cow generically are dealt with as rising stars; specifically we deal with γ Draconis in one case, and Sirius in the other.

The evidence goes to show that these two stars were those to the risings of which very great importance was attached, but they did not stand alone. We get another form of Isis (referring, it is possible, to the star α Columbæ, before even Sirius was used), so that we have a northern star and a southern star observed at the same time—the two eyes of Rā. The other goddesses which have not yet been worked out probably refer to one or other of these stars, or to others which lie more to the south. These are represented rather in the temples above the first cataract than in those below. This fact will be enlarged upon in the sequel.

The study of orientation, then, combined with mythology, supplies us with other rising stars besides Sirius, and, indeed, although the date given by Biot for the first heliacal rising of Sirius at the solstice—3285 B.C.—seems a very remote one, it is practically certain that α Columbæ was previously used, because before that time it was conveniently situated to give warning of the sunrise at the Summer solstice, as Sirius was subsequently. The worship would be kept up after the utility had gone.

Dümichen's view with regard to the local cult of Hathor and its astronomical origin is not very different from mine. He writes :—

"Der Cult der Göttin Hathor geht in die ältesten Zeiten der ägyptischen Geschichte zurück. Schon die Pyramideninschriften erwähnen eine Heliopolitische Hathor und Priester und Priesterinnen dieser Göttin werden in denselben Grabkapellen nicht selten genannt. Die Hathor war keine speciell lokalisirte Gottheit, sondern eine allgemein in sämmtlichen Tempeln Aegyptens verehrte Form eines Cultes, dessen Urgedanke, im weitesten Sinne, die Auffassung des weiblichen Principes gegenüber dem männlichen Principe der Gottheit war. In dieser Auffassung erscheint sie geradezu identisch mit der Isis, weshalb auch beiden Göttinnen die Kuh das geheiligte Thier war. Da in jeder Stadt, vor allen aber in jeder Nomos-Hauptstadt eine Hathor als Schutzgöttin des betreffenden Ortes aufgeführt wird, so ist es erklärlich, dass die lokalen Formen dieser Göttin in den Inschriften der Tempel in grösster Anzahl aufgeführt werden. Im Tempel von Edfu werden Beispiels halber an der Decke des Pronaos über 300 Namen der Göttin mit ihren lokalen Beziehungen hergezählt mit besonderer Bevorzugung derjenigen lokalen Formen, welche in den einzelnen Nomos-Hauptstädten sich eines hervorragenden Cultes erfreuten. Die letzteren berühren vorzüglich eine Sieben-Zahl von Hathoren, welche als die grossen bezeichnet werden und von denen fast in allen grösseren Tempeln Listen an den Wänden zu lesen sind.

"In der älteren Zeit bezeichnet Hathor einen kosmischen Urbegriff. Schon ihr Name verräth aufs Deutlichste die kosmogonische Wurzel. Ha. t. ḥor wörtlich übersetzt "Wohnung des Horus—Behausung Gottes" d. i. die Welt, die Darstellung Gottes in der sichtbaren Welt, die Natur, in welcher die Gottheit wirksam ist."[1]

Before I pass on, it will, I think, be well to point out that the argument I have used to show that Isis was really a generic name is enforced when we consider the allied points relating to Osiris.

It is quite clear that some of the gods symbolised setting stars. We already know that the setting sun became Osiris, Atmu, or Tmu, and, whatever the names, they were all represented as mummies. But the sun was not the only body that was symbolised as Osiris; the moon and stars were at times symbolised in the same way. We may, indeed, venture to make the following generalised statement:—

[1] Dümichen, "Bauurkunde der Tempelanlagen von Dendera," p. 20.

OSIRIS = ANY CELESTIAL BODY BECOMING INVISIBLE.

SUN SETTING.	MOON WANING.	PLANET SETTING.	STARS SETTING.	BODIES PALING AT DAWN.	
				Stars.	Planets.
Osiris	Osiris	Venus as Osiris	Khons-Osiris	Sah-Osiris	Venus
			Ptah-Osiris		Star of Osiris
			Min-Osiris		

It will be observed with what fulness the antithesis of Isis is indicated.

I have already pointed out that the possible temple of Osiris at the pyramids points to the westward, but our special reference now is to stars. When we come to look for this mummy-symbolism among the gods other than sun-gods (it is entirely and remarkably absent among the goddesses), we find Khons, Ptah, and Khem pictured as mummies; that is, they become a sort of Osiris. Supposing that these gods were worshipped, there would probably be temples dedicated to them; still, the absence of such temples would not be decisive, since they might have been destroyed. However, very fortunately for this inquiry, there are two têmples still extant at Thebes, known as the temples of Khons and Ptah. If there is anything, then, in the idea that there must be some relation with the western horizon in the case of these gods represented as mummies, these temples should point to the west. *They do point to the west.*

Very fortunately, also, these temples have a pretty good history: that is, one knows, within some hundreds of years at all events, when they were founded. Therefore, by help of those astronomical methods to which I have previously referred, it is not difficult to get at the stars. They turn out to be a southern star—Canopus—in the case of the temple of Khons, and Capella in the case of the temple of Ptah. Now, there is another very important temple at Thebes, it is a temple without a name, at right angles to the temple of Mut. This also points to the west. Although the evidence is not complete, it clearly suggests that this temple was dedicated to the god Min or Khem, and was oriented to the star Spica; so that at Thebes alone it looks as if the three gods represented by mummies—different stellar forms of Osiris—Khons, Ptah and Min, have all been run to earth in the three stars Canopus, Capella, and Spica.

Provisionally, we may hazard the assertion that the mummy form marks a setting star, as the horns and disk mark a rising one. We get the antithesis between Osiris and Isis.

ISIS, OSIRIS AND HORUS.

We gather, then, that the wonderful old-world myth of Isis and Osiris is astronomical from beginning to end, although Osiris in this case is not the sun, but the moon. But I have not yet finished with the mummy form; the waning moon is also Osiris. It is supposed to be dying from the time of full moon to new moon. The Egyptians in their mythology were nothing if not consistent; the moon was called Osiris from the moment it began to wane, as the sun was Osiris so soon as it began to set. A constellation paling at sunrise was also Osiris.

A "CHANGE OF CULT" AT LUXOR.

I have previously noted the symbolism of Sirius-Hathor as a cow in a boat associated with the constellation of Orion. There is a point connected with this which I did not then refer to, but which is of extreme importance for a complete discussion of the question now occupying us. We get associated with the cow in the boat, Orion (Sah) as Horus, but in other inscriptions we get Orion as a mummy—that is to say, in the course of Egyptian history the same constellation is symbolised as a rising sun at one time and a setting sun at another. Now, that must have been so if the Egyptian mythology were consistent and rested on an astronomical basis, because Sah rose in the dawn in one case and faded at dawn in the other. From the table giving a generalised statement with regard to Osiris, similar to that we have already considered for Isis, it looks as if the mythology connected with Osiris is simply the mythology connected with any celestial body becoming invisible. We have the sun setting, the moon waning, a planet setting, stars setting, constellations fading at dawn. We see, therefore, that the Egyptian mythology was absolutely and completely consistent with the astronomical conditions by which they were surrounded; that, although it is wonderfully poetical, in no case is the poetry allowed to interfere with the strictest and most accurate reference to the astronomical phenomena which had to be dealt with.

The argument, then, for the use of Isis as a generic name is greatly strengthened by the similar way in which the term Osiris, which is acknowledged to be a generic name, is employed.

Now to return to Denderah in the light of the preceding discussion. A curious and interesting thing is that we find that the temple of *Isis*, which is very much ruined, does not contain emblems of the Sirius worship; but that all these appear in the temple of *Hathor*, which, of course, pointing as it does to the north-east, could never have received any light from a star south of the equator. There has been a change of cult.

On the other hand, the temple of Isis presents so many emblems thought to relate to the worship connected with γ Draconis, to which the temple of Hathor was in all probability directed, that it was named the Typhoneum by the French Commission.

There has been an apparent change of *rôle* and cult, due either to the fact that in time the observation of the rising of Sirius superseded that of the rising of γ Draconis, or that the worship of Set was replaced.

With regard to this change of cult, we moderns should have no difficulty. We go to Constantinople and see Mahommedans worshipping in St. Sophia; we go to Greece or Sicily and find Christian worship in many of the old temples. Thus the change of cult in Egypt, which I claim to have demonstrated on astronomical grounds at Denderah, is a thing with which we are perfectly familiar nowadays. The great point, however, is that in Egypt the change of cult might depend upon astronomical change—upon the precession of the equinoxes, as well as upon different schools of religious or astronomical thought. We gather from this an idea of the wonderfully continuous observations which were made by the Egyptians of the risings and settings of stars, because, if the work had not been absolutely continuous, they would certainly never have got the very sharp idea of the facts of precession which they undoubtedly possessed; and it is also, I think, pretty clear that future astronomical study will enable us to write the history of those changes which are now hidden by that tremendous mythological difficulty, which has not yet been faced. That, of course, is not the only difficulty, because the question s clouded by the absence of authentic dates and the perpetual reference to the past which is met with in all the monuments. The Egyptians were much more anxious to bring back to knowledge what happened 1000 years before than to give an idea of the current history of the country.

We have, then, at length arrived at a possible explanation of the difficulties acknowledged in regard to the temples of Denderah in Chapters XIX. and XX.

It is, briefly, that at some epoch observations of the star Sirius replaced, or were added to, those made of γ Draconis. Mythologically, a new Isis would be born.

This point will be referred to later; one of the longest-lasting astro-theological strifes in Egypt was the fight for supremacy between the priests of Amen and the priests of Set. At Denderah the former were ultimately victorious, and hence the change of cult.

This suggestion is based on the following considerations:—

(1) While the Denderah Hathor was represented by the disk and horns on a hippopotamus, at Thebes (the city of the "Bull" Amen) Hathor is represented by a cow with a like head-dress.

(2) Isis, represented originally as a goddess with the two feathers of Amen, standing in a boat, is now changed to a cow with the disk and horns.

(3) Hathor was the "cow of the western hills" of Thebes. It is in these hills that the temple Dêr el-Bahari lies; and this temple, if oriented originally to Sirius, would have been founded about 3000 B.C., when Sirius at rising would have an amplitude of 26° S. of E.

(4) A temple was built or restored later at Denderah, and Sirius with the cow's horns and disk became the great goddess there; and when her supremacy all over Egypt became

undoubted, her birthplace was declared—at Denderah—to have been Denderah.[1]

(5) In the month-list at the Ramesseum the first month is dedicated to Sirius, the third to Hathor. This is not, however, a final argument, because *local* cults may have been in question.

(6) "Set" seems to have been a generic name applied to the northern (? circumpolar) constellations, perhaps because *Set* = darkness, and these stars, being *always visible* in the night, may have in time typified it. Taurt, the hippopotamus, was the wife of Set. The Thigh was the thigh of Set, etc. γ Draconis was associated therefore with Set, and the symbolism for Set-Hathor was the hippopotamus with horns and disk. Now if, as is suggested, Sirius replaced γ Draconis, and the cow replaced the hippopotamus, the cult of Set might be expected to have declined; and as a matter of fact the decline of the worship of Set, which was generally paramount under the earlier dynasties, and even the obliteration of the emblems on the monuments, are among the best-marked cases of the kind found in the inscriptions.

(7) The *Isis* temple of Denderah was certainly oriented to Sirius; the *Hathor* temple was as certainly *not* so oriented. And yet, in the restorations in later times (say, Thothmes III.—Ptolemies), the cult has been made Sirian, and the references are to the star which rises at the rising of the Nile.

So far, then, mythology is with me; but there is a difficulty. According to the orientation theory, the cult must follow the star; this must be held to as far as possible. But

[1] Brugsch thus translates one of the inscriptions:—"Horus in weiblicher Gestalt ist die Fürstin, die Mächtige, die Thronfolgerin und Tochter eines Thronfolger. Ein fliegender Käfer wird (sic?) geboren am Himmel in der uranfänglichen Stadt (Denderah) zur Zeit der Nacht des Kindes in seiner Wiege. Es strahlt die Sonne am Himmel in der Dämmerung, wann ihre Geburt vollbracht wird." Brugsch, "Astron. Inscript.," p. 97.

suppose the precessional movement causes the initial function of a star to become inoperative, must not the cult—which, as we assume, had chiefly to do with the heralding of sunrise at one time of the year or other—change? And if the same cult is conducted in connection with another star, will not the old name probably be retained?

I do not see why the Egyptians should have hesitated to continue the same cult under a different star when they apparently quite naturally changed Orion from a form of Osiris (Sah-Osiris) and a mummy (as he was represented when the light of his stars was quenched at dawn at the rising of Sirius) to that of Sah-Horus (when in later times the constellation itself rose heliacally).

And, moreover, the antagonism of rival priesthoods has to be considered. It is extremely probable that the change of a Set temple at Denderah into a Theban Hathor-temple was only one example of a system generally adopted, at least in later times.

CHAPTER XXX.

THE TEMPLE-STARS.

THE two preceding chapters should have suggested that if there be any truth in the astronomical and mythological views therein put forth, there should be other stars to deal with besides Sirius and γ Draconis, and other temples besides those at Annu, Denderah and Thebes which have to be studied.

This is so, and I now propose to give a general account of the conclusions so far arrived at, but I must *in limine* state that the account must be a brief one and more suggestive than final, for the reason that the lack of accurate local data stops the way.

In an inquiry of this kind it is well to work slowly out from the known. The facts which have been given will, I think, cause it to be generally agreed that in the temple of Isis at Denderah we have a structure which the inscriptions, as well as astronomical inquiry, show was certainly a temple oriented to Sirius. The other fact that New Year's Day in the Nile valley was determined for thousands of years by the heliacal rising of that star, is among the most familiar in the domain of Egyptology.

Obviously, then, the first inquiry must refer to the possible existence of other Sirian temples.

From 3285 B.C., when Sirius rose heliacally at the solstice, its declination has varied from 24° S. to 16½° S. in 500 B.C. The corresponding amplitudes for Thebes being 26½° and 18° S. of E.

Between these amplitudes we find the following temples :—

SIRIUS.

Place and Temple.	Amp.	Sea Horizon.		Hills 1° High.		Hills 2° High.		Remarks.
	S. of E.	Dec. S.	Years.	Dec. S.	Years.	Dec. S.	Years.	
Karnak (Temple O) - (Gr.)	26°½	24°	3300	23½°	**3150**	23°	3050	This may have been a solar temple, as its amplitude is nearly equal to that of the sun at the winter solstice.
Dêr el-Bahari (Gr.)	24½°	22¼°	2850	21¾°	**2700**	21¼°	2575	
Dosche	21½°	20¼°	2225	19¾°	**2050**	19½°	2000	
Karnak (Temple D) (Gr.)	21½°	19½°	2000	19°	**1800**	18½°	1600	
Naga (Temple G) (Gr.)	19°	18¼°	1500	18°	**1400**	17¾°	1250	
Philæ (Ethiopian Temple)	19½°	18°	1400	17½°	1100	17°	**800**	Hills at least 2° high.
Denderah (N.W. Temple)	18½°	16¾°	**700**					Hills very low.

It is quite clear that we must not look for Sirian temples before 3200 B.C., because the heliacal rising of Sirius at Thebes before that time did not take place near the solstice. The above table shows that the earliest Sirian temple really dates from about 3000 B.C.[1]

But what star did Sirius replace? An inspection of a precessional globe shows at once that the star which rose heliacally at the solstice before Sirius was α Columbæ (Phact). Its declination has varied from 57° S. at 5000 B.C. to 37° S. at 0.

We have the following temples which might have been oriented to this star; and here I must repeat that once a star has been symbolised as a god or a goddess on account of its astronomical utility, the cult would be continued after the utility had ceased—that is, in this case, after Sirius had replaced Phact astronomically.

[1] In this and the following tables the dates connected with the heights of hills where they are known are given in heavy type. Where the local conditions are unknown, hills 1° high have been assumed.

PHACT.

Place and Temple.	Amp.	Sea Horizon.		Hills 1° High.		Hills 2° High.		Remarks.
	S. of E.	Dec. S.	Years.	Dec. S	Years.	Dec. S.	Years.	
Memnonia (Western Temple)	58½°	50½°	3750	49¾°	**3700**	49°	3550	Hills low.
Barkal (Temple B).	53½°	50°	3250	49¼°	**3600**	48¾°	3500	
Karnak V	56½°	49°	3550	48¼°	**3400**	47½°	3250	
Abu Simbel (Hathor Temple)	54°	48¾°	3500	48°	3350	47½°	**3250**	Hills nearly 2° high.
Dêr el-Medinet (Gr.)	54½°	47½°	3250	46¾°	**3050**	46°	2900	
Saboa	51¼°	46°	2900	45½°	**2750**	45°	2650	
Karnak (Temple J) (Gr.)	51½°	45¼°	2700	44½°	**2525**	43¾°	2300	
Medînet Habû (Small J J) (Gr.)	51½°	45¼°	2700	44½°	**2525**	43¾°	2300	
Barkal (Temples J and H)	47½°	44½°	2525	44°	**2400**	43½°	2250	
Surarieh	51°	43½°	2250	42¾°	**2050**	42°	1850	
Medînet Habû (Palace K K) (Gr.)	46½°	40¾°	1500	40°	**1250**	39½°	1050	
Medînet Habû (Ethiopian Temple)	45°	40°	1250	39°	**900**	38½°	**500**	The hills may be taken as a little over 1° high.

The temple of Hathor at Abu Simbel, embellished by Rameses II., was in all probability a shrine dedicated to Amen-t-Hathor about 3200 B.C. Amen-t seems to have been an Ethiopian goddess, for we hear nothing of her at Heliopolis or Memphis.

It follows that if this be so, Sirius succeeded to α Columbæ precisely as γ Draconis succeeded to Dubhe; but temples could still be dedicated to the old Hathor α Columbæ, while this was not possible for Dubhe, because it became circumpolar and never rose.

It may also be pointed out that the temple V of Lepsius at Karnak finds its place in a series by supposing it to have been

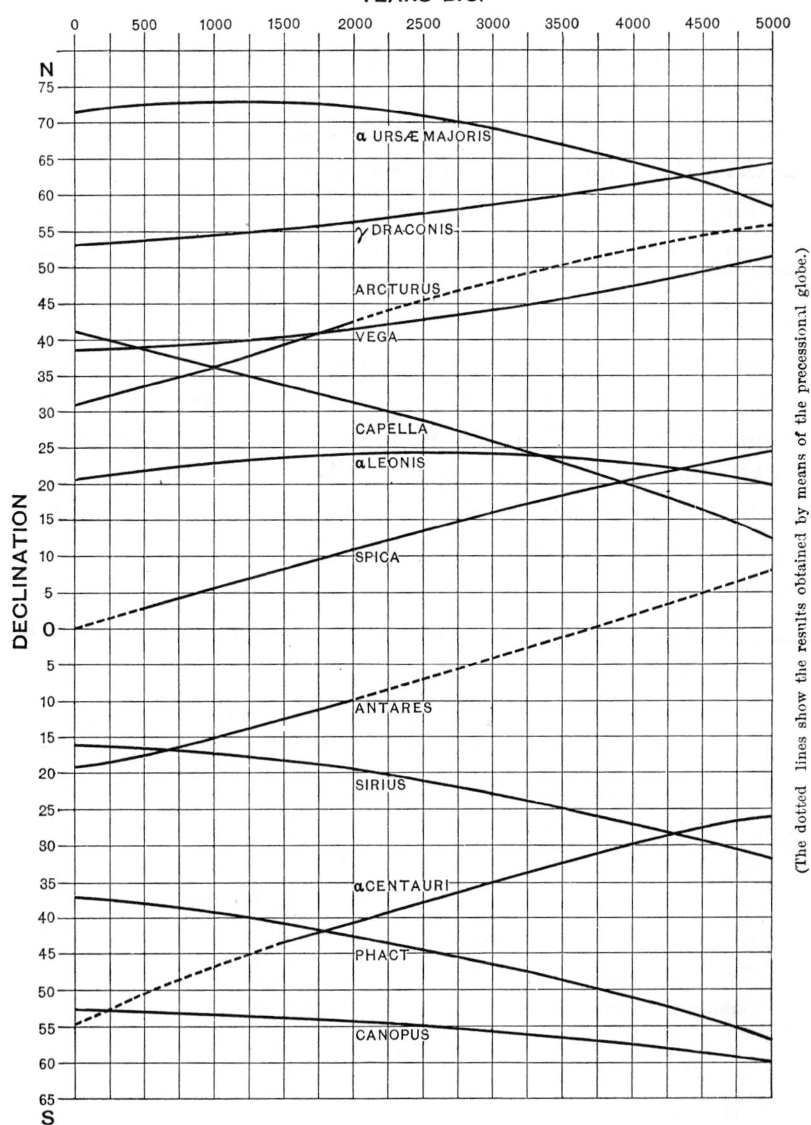

CURVES SHOWING THE DECLINATIONS OF SOME OF THE STARS USED BY THE EGYPTIAN ASTRONOMERS AT DIFFERENT EPOCHS.

oriented to the S.E. instead of the N.W. as shown in Lepsius' maps. Such a mistake might easily have arisen in consequence of its ruined condition. It may be stated in favour of my view that I am acquainted with no temple in Egypt directed between the amplitudes 35° and 90° N. of W.

But so far we have dealt only with the summer solstice, and yet in Egypt there were people who lived in towns with E. and W. walls who, I take it, must have had a worship depending upon the equinoxes.

About 3500 B.C., Antares (α Scorpii) rose heliacally at the autumnal equinox as α Columbæ did, as we have seen, at the summer solstice. There is not much doubt, from the symbol of Serk-t that this goddess represented a star in the Scorpion. Further, at that date its rising took place due east, so any E. and W. temple—and many existed in *Lower* Egypt—might have been then used for observations of this star.

But about the same time the southern star, α Centauri, could have been used to herald the sunrise at the autumnal equinox.

α CENTAURI.

Place and Temple.	Amp.	Sea Horizon.		Hills 1° High.		Hills 2° High.		Remarks.
	S. of E.	Dec. S.	Years.	Dec. S.	Years.	Dec. S.	Years.	
Barkal E	$33\frac{1}{2}$°	$31\frac{3}{4}$°	3625	$31\frac{1}{4}$°	**3700**	$30\frac{3}{4}$°	3800	
Kûrnah (Seti I.)	$35\frac{1}{2}$°	$31\frac{3}{4}$°	3625	$31\frac{1}{4}$°	**3700**	$30\frac{3}{4}$°	3800	Hills low.
Kûrnah (Palace)	36°	$32\frac{1}{4}$°	3500	$31\frac{3}{4}$°	**3625**	31°	3750	Hills low.
Wady Halfa (Thothmes II.)	$38\frac{3}{4}$°	$35\frac{3}{4}$°	2900	$35\frac{1}{4}$°	**3000**	$34\frac{3}{4}$°	3075	
Barkal L	38°	36°	2850	$35\frac{1}{2}$°	**2950**	35°	3030	
Wady Halfa (Thothmes III.)	40°	$36\frac{3}{4}$°	2725	$36\frac{1}{4}$°	**2800**	$35\frac{3}{4}$°	2900	
Wady E. Sofra	$38\frac{1}{2}$°	37°	2675	$36\frac{3}{4}$°	**2700**	$36\frac{1}{4}$°	2800	
Memnonia Rameses II. (Mean of Fr. & Gr.)	43°	$38\frac{1}{4}$°	2475	$37\frac{1}{2}$°	**2600**	37°	2700	Hills low.
Kom Ombo (Little Temple)	$43\frac{1}{2}$°	39°	2375	$38\frac{1}{2}$°	**2450**	$37\frac{3}{4}$°	2575	

It would appear that several temples were directed to this star in *Upper* Egypt from 3700 B.C. onward. The series of them is shown in the preceding table.

For the vernal equinox, so far, I have found no temples besides those directed due E. in which the rising of the Pleiades may have been watched. It is more than probable that the worship of the sacred bull by the Memphitic inhabitants of Egypt may have been connected with this constellation. Certainly in pyramid times Neith and Serk-t were both worshipped, and the goddesses under whose protection the Canopic vases were supposed to be—Isis, Nephthys, Neith, and Serk-t—may have symbolised the two solstices and the two equinoxes.

We may next consider the complete series of N.E. temples represented at Heliopolis, Denderah and Thebes. These we must, as I have shown in Chapter XX., divide into two series, dealing with α Ursæ Majoris before it became circumpolar, and γ Draconis afterwards.

I have already (p. 203) stated that α Lyræ may possibly have preceded both α Ursæ Majoris and γ Draconis as a representative of Set, but no table is necessary.

The first series, dealing with α Ursæ Majoris, is as follows:—

α URSÆ MAJORIS.

Place and Temple.	Amp.	Sea Horizon.		Hills 1° High.		Hills 2° High.		Remarks.
	N. of E.	Dec. N.	Years.	Dec. N.	Years.	Dec. N.	Years.	
Annu	77°	57°	**5200**	58°	5050	59°	4900	Hills low.
Denderah	$71\frac{1}{2}$°	$57\frac{3}{4}$°	5100	$58\frac{3}{4}$°	4950	$59\frac{3}{4}$°	**4800**	Hills 2° high.
Denderah	78°	$60\frac{3}{4}$°	4600	62°	4400	63°	**4200**	,, ,, ,,

The second series, dealing with γ Draconis, is naturally much fuller.

γ DRACONIS.

Place and Temple.	Amp.	Sea Horizon.		Hills 1° High.		Hills 2° High.		Remarks.
	N. of E.	Dec. N.	Years.	Dec. N.	Years.	Dec. N.	Years.	
Redesieh	77½°	61¾°	4250	62¾°	**4600**	63¾°	4850	
Karnak (Z and X)	72½°	58¾°	3100	59¾°	**3500**	60¾°	3900	
Dakkeh	69¼°	58¾°	3100	59¾°	**3500**	60½°	3800	
Denderah	71½°	57¾°	2650	58¾°	3100	59¾°	**3500**	Hills 2° high.
Annu (Restoration)	77°	57°	**2300**	58°	2800	59°	3200	
Karnak W	68½°	56½°	2100	57½°	**2550**	58¼°	2900	
Karnak [1]A M C	63½°	53¼°	300	54¼°	1000	55°	1400	Hills 1½° high. 54½°. **1200.**

The table brings before us the remarkable fact that at Redesieh and Denderah, which both lie on the two old roads from the Red Sea into Upper Egypt, we have the first traces of the worship of Set: in other words, of observations during the night in that region, as we found it at Annu.

As *a* Ursæ Majoris and γ Draconis were observed in the extreme north, so several stars appear to have been observed near the south point, among them Canopus (*α* Argûs), towards which star the temples shown in the following table seem to have been directed, among them the well-known temple of Khons at Karnak, so that provisionally we may take that divinity as a personification of the star. Granting this, it will be noted that the introduction of this cult into Thebes was late; this is quite in harmony with the statements of Egyptologists, who point out that this god has the side-lock, indicating youth, and that he was the latest addition to the Theban Triad.

In later times the curve of declination of this star is so flat that most accurate measures are required.

[1] With regard to the temple A of Lepsius, it may be stated that in the time of the Ptolemies it received considerable and curious enlargements and embellishments which make it unique among Egyptian temples.

CANOPUS.

Place and Temple.	Amp.	Sea Horizon.		Hills 1° High.		Hills 2° High.		Remarks.
	S. of W.	Dec. S.	Years.	Dec. S.	Years.	Dec. S.	Years.	
Karnak B	63½°	54¼°	**2150**	53¼°	1300			Hills 1½° high give us 1800 B.C.
Naga (*f*)	57°	53¾°	1700	53¼°	**1300**	52¾°	300	Hills 1½° high give us **1400** B.C.
Karnak (Seti II.)	63°	53¾°	1700	53°	1000			
Karnak (Khons)	62°	53°	1000	52¼°	300 A.D.			Hills 1½° high give us **300** B.C.

When we attempt to trace the *most* southerly stars to which temples were erected in Upper Egypt, we find a series of temples which are very remarkable in several respects from the orientation point of view. Their amplitudes are all above 74°, one being as high as 86½.° They all face South of *West*, and when their latitudes are taken into account, the very striking thing comes out that the declination of the star observed was very nearly the same—that is, that probably *all the temples were founded at about the same time to observe the same star*.

The facts are as follows:—

Temple.	Amplitude S. of W.	Declination S.	
Edfû	86½° ..	64¾° ..	Hills 1° high.
Philæ Y	76½° ..	64° ..	Hills 2° high.
Semneh	76½° ..	64¾° ..	Local conditions not known.
Amada	74½° ..	64¼° ..	

With regard to the Philæ temple, the amplitude is uncertain, as the measures do not agree; but if we reject Philæ the other coincidences are too remarkable to be neglected.

It is to be hoped that a complete survey of the island will soon be undertaken.

Now, I cannot find any important stars to fit this declination since 7000 B.C. except Canopus and Phact, and the latter is barred because it was used as a *rising star*, and indeed was the first solstitial Isis.

If we inquire into the conditions relative to Canopus, we find that star had the declination of 64° about 6400 B.C., and that, as determined by the precessional globe, it then set heliacally at the autumnal equinox.

If we assume that Canopus is in question, the break between the dates 6400 B.C. and 2150 B.C. has to be explained. There may have been temples at Thebes now destroyed. There seems no doubt that the temple at Philæ, lettered Y by the French and L in Baedeker's Handbook, was the most ancient one on the island, and that the cult was similar to that at Edfû.[1]

It will be most interesting to see whether the suggestion that Canopus was observed in early times at Philæ and Edfû especially, be confirmed.

It is clear that for these and other southern temples an examination of the local conditions and a determination of the places of the southern stars are necessary before the other southern gods and goddesses can be worked out.

We next come to the N.W. quadrant. Here, apparently, we have only to deal with Capella and Spica. Summarising the information detailed in a previous chapter, we find the following temples probably erected to these stars:—

CAPELLA.

Place and Temple.	Amp.	Sea Horizon.		Hills 1° High.		Hills 2° High.		Remarks.
	N. of W.	Dec. N.	Years.	Dec. N.	Years.	Dec. N.	Years.	
Memphis	12°	10°	5500	$10\frac{3}{4}$°	**5350**	$11\frac{1}{4}$°	5300	
Annu	13°	11°	**5325**	$11\frac{1}{2}$°	5250	12°	5200	
Karnak U	$27\frac{1}{2}$°	$24\frac{1}{4}$°	3250	$24\frac{3}{4}$°	3150	$25\frac{1}{4}$°	**3050**	
Thebes (Petit Temple du Sud)	$31\frac{1}{2}$°	$27\frac{3}{4}$°	2600	$28\frac{1}{2}$°	2500	29°	**2400**	
Karnak G	35°	$30\frac{3}{4}$°	2050	$31\frac{1}{2}$°	1925	32°	1850	$32\frac{1}{2}$°. **1750.** Hills 3° high

[1] Baedeker, "Ober-Aegypten," p. 320.

SPICA.

Place and Temple.	Amp.	Sea Horizon.		Hills 1° High.		Hills 2° High.		Remarks.
	N. of W.	Dec. N.	Years.	Dec. N.	Years.	Dec. N.	Years.	
Karnak Y	17½°	15½°	2850	16°	2950	16½°	**3050**	17°. **3200.** Hills 3°.
Tell el-Amarna	13°	10¾°	**1900**					

The temples oriented to Capella and Spica are discussed in the next chapter.

The information given in the present chapter may be completed by a table showing the warning stars available for heralding sunrise about the times when the orientations suggest that the various temples were originally founded.

To prepare this table I have used the precessional globe previously referred to. The results are therefore rough, as the ecliptic has been taken as fixed; but they are useful for the purpose of a reconnaissance. The table shows the stars on the horizon, or near it, at the equinoxes and solstices when the sun was 10° below the horizon. When the star was not exactly on the horizon when the sun was 10° below, its position above or below at that moment is indicated in the table by giving the number of degrees the star was above (+) the horizon or below it (–) at the time.

The dates taken are those most conveniently given by the globe, being those in fact occupied by the pole of the equator at some one or other of twenty-four equidistant points on a circle round the pole of the ecliptic starting from 1880 A.D. as zero.

It will be seen that all the stars referred to in the preceding tables occupied positions of great importance between 6000 B.C. and 2500 B.C., and that there are several southern stars indicated which eventually may be useful in the discussion of the southern temples.

TABLE OF HELIACAL RISINGS AND SETTINGS.

Date B.C.	Spring Equinox.		Summer Solstice.		Autumnal Equinox.		Winter Solstice.	
	Rising.	Setting.	Rising.	Setting.	Rising.	Setting.	Rising.	Setting.
5675			Phact		Vega. −1	Canopus	α Phenicis	β Muscæ
4600	Aldebaran		Phact		γ Draconis αTrianguli	Capella Canopus +3		
3525	αPhenicis Pleiades +3	Antares +2	Phact		Antares −1 α Centauri +3½	Aldebaran +2		
[3200]			Sirius					
2450		α Pavonis	Sirius +3	Altair	β Argus +3			

The real precedence of Capella and Spica in temple-building is not shown in the above table, because these stars were not used either at the solstices or the equinoxes.

CHAPTER XXXI.

THE HISTORY OF SUN-WORSHIP AT ANNU AND THEBES.

Now that we have been able to discuss with more or less fulness the stars—very few in number—to which the temples in both Upper and Lower Egypt were probably oriented, and further, the astronomical requirements which they were intended to fulfil, we are in a position to consider several questions of great interest in relation to the earliest observations of the sun and stars.

One of the first among these questions is whether the complete inquiry throws any light upon the suggestion made on page 85, that in different temples we seem to be dealing with at least two different kinds of astronomical thought and methods; as if, indeed, we were in presence of ideas so differently based that the assumption of different races of men, rather than different astronomical and religious ideas, is almost necessary to account for them.

Let us begin with the apparent result of the inquiry into sun-worship as practised at Annu and Thebes.

It was suggested that, although in the matter of simple worship the sun would come before the stars, in *temple*-worship the conditions would be reversed in consequence of the stable rising- and setting-places of the latter as compared with those of the sun at different times of the year.

Another suggestion was hazarded that sun temple-worship might have been an accidental result of the sunlight entering a temple which had really been built to observe a star; and that such temple sun-worship might possibly have preceded the time at which the solstices and equinoxes, and their importance, had been made out. I think it is possible to show that this really happened, and we owe the demonstration of this important fact to the Egyptian habit of having two associated temples at right angles to each other, because this habit justifies the assumption that at Annu the mounds and single obelisk which now remain not only indicate the certain existence in former times of one temple, but, in all probability, of two at right angles to each other.

The next question we have to consider is whether the researches at Annu bear this surmise out. Let me refer to what has already been stated. As I have shown in Chapter VIII. (p. 77), the north and south faces bear 13° north of west —13° south of east. I have elsewhere shown (Chap. XXI., p. 215) that there is good reason for believing that the original foundation of the temple at Annu dates from the time when the north-pointing member of such a double system was directed to α Ursæ Majoris. This was somewhat earlier than 5000 B.C.

Bearing in mind the facts obtained with regard to other similar rectangular systems, we are led to inquire whether at that date a temple oriented to declination 11° north, that is the declination proper to the amplitude of the member looking west, was directed to any star.

We find that the important star Capella was in question.

Now, so far in my references to stars, little mention has been made of Capella. It is obvious that the first thing to be done on the orientation hypothesis is to see whether any other temple—and if of known cult, so much the better—is found oriented to Capella. There is one such temple; it was erected by Thothmes III. (Time of Thothmes, 1600 B.C. Amplitude of temple, 35° west of north = with hills 3° high, $32\frac{1}{2}$°

north declination; Capella 33° north declination about 1700 B.C.) It is the temple of *Ptah* at Karnak.

And now it appears there is another. During the year 1892 the officers of the Museum of Gîzeh, under the direction of M. de Morgan, excavated a temple at Memphis to the north of the hut containing the recumbent statue of Rameses, and during their work they found two magnificent statues of Ptah, "les plus remarquables statues divines qu'on ait encore trouvées en Égypte,"[1] and a colossal model in rose granite of the sacred boat of Ptah.

These discoveries have led the officers in question to the conclusion that the building among the ruins of which these priceless treasures have been found is veritably the world-renowned temple of Ptah of Memphis. It may, therefore, be accepted as such for the purpose of the present inquiry, although it is difficult to reconcile its *emplacement* in relation to the statues with the accounts given by the Arab historians.

In January, 1893, Captain Lyons, R.E., was good enough to accompany me to determine the orientation of the newly uncovered temple walls. We had already, two years previously, carefully measured the bearings of the statues of Rameses. We found the temple in all probability facing westwards, and not eastwards; this we determined by a seated statue facing westwards; and we concluded its orientation, assuming a magnetic variation of $4\frac{1}{2}°$ west, to be $12\frac{3}{4}°$ north of west, and the hills in front of it, assuming the village of Mît-Rahîneh non-existent, to be 50′ high.

Here, then, we get reproduced almost absolutely the conditions of the obelisk at Heliopolis in a Ptah temple oriented to Capella 5200 B.C.

We are driven, then, to the conclusion that the star Capella

[1] New Gîzeh Catalogue, p. 61.

is personified by *Ptah*, and that as Capella was worshipped setting, Ptah is represented as a mummy. If this be so, we must also accept another conclusion: the temples both at Annu and Memphis were dedicated to Ptah.

About 5300 B.C. we seem almost in the time of the divine dynasties, and begin to understand how it is that in the old traditions Ptah precedes Rā and is called "the father of the beginnings, and the creator of the egg of the Sun and Moon."[1]

We are driven to the conclusion that this worship at Annu and Memphis was the worship of the sun's disc when setting, at the time of the year heralded by Capella, when it had the declination of 10° north. The dates on which the sun had this declination were, as already stated, about April 18 and August 24 of our Gregorian year. The former, in Egypt, dominated by the Nile, was about the time of the associated spring and harvest festivals.

So much for the Ptah mummy form of the Sun-God, to which the *Theban* priests erected no important temples. There was still another mummy form of the Sun-God, the worship of which existed at Thebes, but which they did their best to abolish by the intensification of the worship of Amen-Rā.

At Thebes, as we have seen, the temple of Mut is associated with one at right angles to it, facing north-west. The amplitudes are $72\frac{1}{2}°$ north of east and $17\frac{1}{2}°$ north of west. I have shown that the temple of Mut would allow γ Draconis to be seen along its axis about 3200 B.C. *I now state that* Spica *would be seen along the axis of the rectangular temple at the same time.*

The cult in this temple-system there can be no doubt, I

[1] Brugsch, "Religion und Mythologie," p. 111. Pierret, "Salle Historique de la Galerie Égyptienne" (du Louvre), p. 199.

think, was the worship of Min, otherwise read Amsu, or Khem in ithyphallic mummy form. This was associated possibly with a harvest-home festival on May 1. (Amplitude of temple, $17\frac{1}{2}°$ north of west = declination 15° = sun's N. declination on May 1.)

Both at Annu and Thebes, therefore, before the temple of Amen-Rā at the latter place became of importance, the sun was worshipped in a temple pointed neither to a solstice nor to an equinox.

It seems, then, that the suggestion that *possibly* sun-worship existed before any great development of the solstitial solar worship is amply justified.

We have next to consider what had taken place at Thebes, so far as we can trace it on the orientation hypothesis after 3200 B.C., when apparently the Spica temple and the associated Mut temple were founded.

To do this it is important to study the masterly essay by M. Virey, entitled "Notices Générales," on the discoveries made at Dêr el-Bahari by MM. Maspero and Grébaut, which is to be found in the new edition of the Gîzeh Catalogue.[1] M. Virey makes us acquainted with the politics of the Theban priests, or rather of the confraternity of Amen which they had founded.

From his account of the confraternity and of the various attempts made by it to acquire political power, however, we gather that it was not only intended to intensify the cult of Amen-Rā at the expense of the sun-worship previously existing at Thebes (in the Spica temple), but that one of the chief aims of the confraternity of Amen was to abolish the worship of Set, Sit, Sut, or Sutech; that is, as I think I have

[1] "Notices des Principaux Monuments Exposés au Musée de Gîzeh," p. 260. (1893.)

proved, generically, the stars near the North Pole, and, as it can be shown, in favour of the southern ones.

The temple of Mut was the chief temple at Karnak in which the cult of the northern stars was carried on, and this was associated with the Spica temple; so both these temples had to go.

We can now realise what the Theban priests got Thothmes to do. They were strong measures, since in his day the cult of Spica (the solar disc, Aten, Min, Khem), and γ Draconis (the Hippopotamus-and-Lion Isis) was supreme.

The little shrine of the Theban Amen was enlarged and built right across the fair-way of the temple of Mut, so that the worship was as effectively stopped as the worship of Isis (when it was prohibited by law) was stopped at Pompeii by the town authorities bricking up the window by which the star was observed.[1]

Further, the shrine so restored was to be of such magnificence that the Spica temple, which had hitherto held first rank, became an insignificant chapel in comparison. Nor was this all: in order still to emphasise the supremacy of Amen-Rā, a third-rate temple was erected to Ptah.

It is clear from this that we must date the great supremacy of the cult of Amen-Rā in and after the time of Thothmes III., and that the cult superseded at Thebes was largely based upon the old worship at Annu.

Now, one of the most remarkable events in Egyptian history was the so-called apostasy of Amen-hetep IV., some hundred and fifty years after Thothmes III.

[1] The little temple of Isis at Pompeii and the associated frescoes in the Naples Museum are well worth careful study, especially with regard to the arrangements made for the stellar observations (and their final stoppage by the drastic proceeding referred to in the text), and the evolution of Horus in Greek times. The Hippopotami are most carefully drawn.

In the time of Thothmes III. the alliance between the royal and the sacerdotal power was of the closest, and in no time of the world's history have priests been more richly endowed than were then the priests of Amen. Not content, however, with their sacred functions, they aimed at political power so obviously that Thothmes IV. and Amen-hetep III., to check their intentions, favoured the cults and priesthoods of Annu and other cities of the north. Amen-hetep III. and his son, Amen-hetep IV., also looked for alliances out of Egypt altogether, and entered into diplomatic relations with the princes of Asia, including even the king of Babylon. This brought him and the priests to open warfare. He replied to their anger by proscribing the cult of Amen, and the name of Amen was effaced from the monuments; still the priestly party was strong enough to make it unpleasant for the king in Thebes; and, to deal them yet another blow, he quitted that city and settled at Tell el-Amarna, at the same time, according to the statement of M. Virey, reviving an old Heliopolitan cult. He took for divine protection the solar disk *Aten*, "which was one of the most ancient forms of one of the most ancient gods of Egypt, Rā of Heliopolis."[1] Now let us say that the time of Amen-hetep IV., according to the received authorities, was about 1450 B.C. The lines of the "Temple of the Sun" at Tell el-Amarna are to be gathered from Lepsius' map, reproduced in the illustration on the next page. The orientation is 13° north of west.[2] This gives us a declination of 11° north, and the star Spica at its setting would be visible in the temple.

Still the light would not enter it *axially* if the orientation is correct. This would have happened in 2000 B.C., that is,

[1] Gîzeh Catalogue, 1893, p. 63.

[2] Professor Flinders Petrie has been good enough to send me his recent measurements. They justify those obtained from Lepsius' plan.

600 years before the time of Amen-hetep IV. This is a point which Egyptologists must discuss; it is quite certain that such a pair of temples as those of which Lepsius gives us the

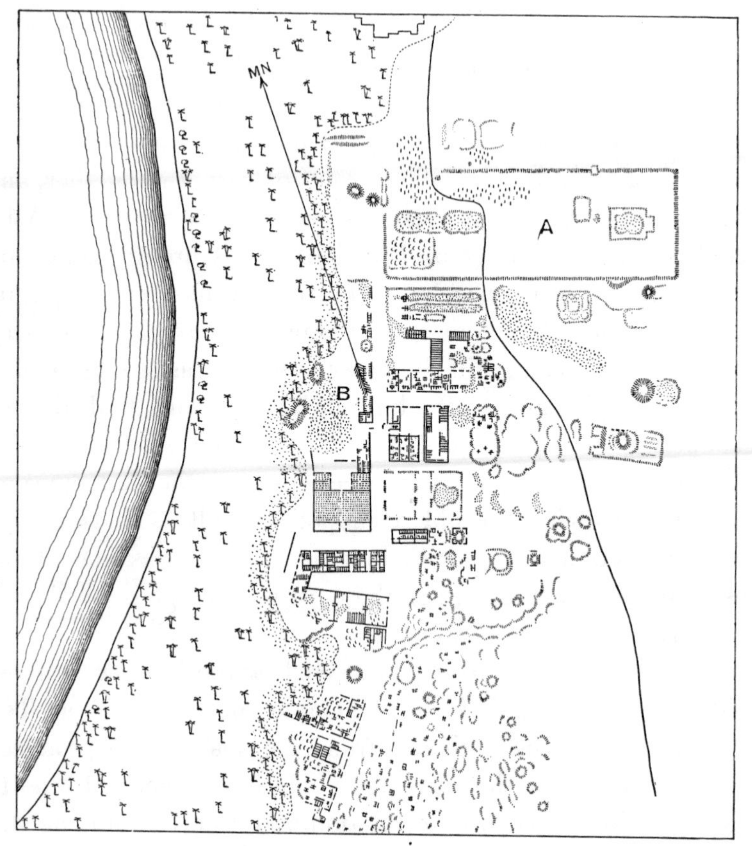

THE TEMPLES AT TELL EL-AMARNA. A, The Aten (Spica) Temple; B, the Set Temple.

plans could not have been completely built in his short reign, and they would not perhaps have been commenced on *heretical* lines in any previous reign during the Eighteenth dynasty. They must therefore have been commenced before 1700 B.C., perhaps in the Seventeenth dynasty. In any case they were certainly finished by Khu-en-Aten.

Professor Flinders Petrie has been good enough, in reply to an inquiry, to state his opinion that the temple was entirely built by Khu-en-Aten. Should this be confirmed, it may have been oriented directly to the sun, on the day named, or was probably built parallel to some former temple, for traces of other temples are shown on Lepsius' plan, and I presume Khu-en-Aten is not supposed to have built all of them.

What, then, was this worship which had been absent from Thebes, but which had held its own to the north to such an extent that Amen-hetep IV. went back to it so eagerly? It could not have been the worship of Capella as a star alone, for such worship had been provided for by Thothmes III. by building temple G. Nor could it have been the worship of Spica as a star alone, for in that case the precedent of Annu would not have been appealed to.

The worship he emphasised there exactly resembled that which had in early times been paramount at Heliopolis. One based on it, but not identical with it, had been in vogue at Thebes from 3200 B.C. to the time of Thothmes III., who, as the tool of the confraternity of Amen, intensified the solstitial worship, and did his best to kill that which had been based upon the Heliopolis cult.

I say *exactly resembled*, because Amen-hetep IV., or some one of the preceding kings of Egypt, when reintroducing the old worship at Tell el-Amarna, orients the solar temple 13° north of west according to the data available. Now when we take the difference of latitude between Heliopolis and Tell el-Amarna into account, we find that the same declination (within half a degree) is obtained from both.

Hence, at Annu in the old days, and at Tell el-Amarna

afterwards, the sun was worshipped on the same day of the year. At both places the sunlight at sunset would enter the temple on April 18 and August 24 of the Gregorian year; hence both temples were probably built really to observe the sunset on a special day. In this view how appropriate was the prayer of Aāhmes, Khu-en-Aten's chief official—

> "Beautiful is thy setting, thou sun's disk of life, thou Lord of Lords and King of the worlds. When thou unitest thyself with the heaven at thy setting, mortals rejoice before thy countenance and give honour to him who has created them, and pray before him who has formed them, before the glance of thy son who loves thee the King Khu-en-aten. The whole land of Egypt and all peoples repeat all thy names at thy rising, to magnify thy rising in like manner as thy setting."[1]

As may be gathered from Lepsius' maps and plans, this "temple of the Sun" was not built alone. Set was again brought to the front. There was another at right angles to it, and while Spica was seen setting in one, a star near γ Draconis was rising in the other.

It may be added that it was not apparently till Rameses II. built his temple M that Set again had an available temple at Karnak: one, however, again to be blocked when the victorious Tirhaqa and the Theban priests returned after their exile. (See page 186.)

We see, then, that in a detailed study of the sun-worship at Thebes alone, we distinctly trace two schools of astronomical thought associated with different religious tendencies. As a protest against the Southern worship of the Theban priests, Khu-en-Aten goes back to a Northern cult. This point is evidently worth further inquiry.

[1] Brugsch, "Egypt," 1891, p. 220.

CHAPTER XXXII.

THE EARLY TEMPLE AND GREAT PYRAMID BUILDERS.

IN previous chapters I have referred to the difference in astronomical thought evidenced by the solstitial solar worship at Thebes as opposed to the non-solstitial solar worship at Annu, and again by the observations of southern stars above Thebes as opposed to observations of high northern stars below.

There is still another fundamental difference to be signalised, and that is the building in some cases of pyramids, with or without associated temples, east and west true.

It will perhaps be generally conceded that the differences in thought indicated by the building or non-building of colossal pyramids are greater than those indicated by the two other *differentia* to which I have referred, and on this ground I propose to enter upon this point at some length.

We may first inquire if there be any other class of considerations which can be utilised to continue the discussion of the question thus raised on astronomical grounds. It is obvious that if sufficient tradition exists to permit us to associate the different classes of structures which have been studied astronomically with definite periods of Egyptian history, a study of the larger outlines of that history will enable us to determine whether or not the critical changes in dynasties and rulers were or were not associated with critical changes in astronomical ideas as revealed by changes in temple-worship and pyramid building. If there be no connection the changes may have been due to a change of idea only—a variation in astronomical thought—and the suggestion of a distinction of race falls to the ground.

In a region of inquiry where the facts are so few and difficult to recognise among a mass of myths and traditions, to say nothing of contradictory assertions by different authors in their exposition of the inscriptions, the more closely we adhere to a rigidly scientific method of inquiry the better. I propose to show, therefore, that there is one working hypothesis which seems to include a great many of the facts, and I hope to give the hypothesis and the facts in such a way that if there be anything inaccurately or incompletely stated it will be easy at once to change the front of the inquiry and proceed along the new line indicated.

I may begin by remarking that it is fundamental for the hypothesis, that the temple of Annu or Heliopolis existed, as stated by Maspero and other high authorities, before the times of Mini (Mena) and the pyramid builders.

Before Mini, according to Maspero, "On et les villes du Nord avaient eu la part principale dans le développement de la civilisation Égyptienne. Les prières et les hymnes, qui formèrent plus tard le noyau des livres sacrés, avaient été rédigés à An."

My observations of the orientation of the obelisk at Annu show that the temple of which it formed part may have possibly been an early member of the series which includes the temple of Mut at Thebes, and of Hathor at Denderah; that is, the worship of Set was in question, to speak generically. Now, according to Maspero, Sit or Set formed one of the divine dynasties, being associated with the sun and air gods at Annu, *i.e.* with Rā, Atmu, Osiris, Horus, and Shou.

It is also certain that the solar temple at Annu at right-angles to the Sit temple, was pointed north-west, and probably to Capella setting, about 5000 B.C.

So much for the astronomical antiquity of Annu. But there are other northern towns besides Annu for which a very high antiquity is claimed.

On this point here is the opinion of Ebers and Dümichen, two of our highest authorities: " Das ist die älteste Stadt in Aegypten, und das mit ihm verbundene Abydos kann nicht viel junger gewesen sein, denn schon im alten Reiche wird es vielfach als heilige Stadt erwähnt."[1]

The sacred character of Abydos is also pointed out by Maspero.[2]

" C'est comme ville sainte qu'elle était universellement connue. Ses sanctuaires etaient célèbres, son dieu Osiris vénéré, ses fêtes suivies par toute l'Égypte; les gens riches des autres nomes tenaient à honneur de se faire dresser une stèle dans son temple."

If it be found that the references to "ancestors," and "divine ancestors," occur after the eleventh dynasty, the race represented by Annu, or the one which immediately followed it (? the Hor-Shesu) may be referred to (*see* the chapters on the Egyptian year).

Of Abydos astronomically I can only say very little, as the various statements as to the orientation of the north-east temples there by various authors are so conflicting that nothing certain can be made out. As they stand they are suggestive that these temples may possibly be associated with that at Luxor, and it may be gathered from the description of them by Ebers and Dümichen in Baedeker

[1] Ebers and Dümichen, in Baedeker's "Ober-Aegypten," p. 59.

[2] *Op. cit.*, p. 21.

that many references to Set (Anubis) occur in the inscriptions. If subsequent measurements indicate that Abydos and Luxor are to be treated together, then astronomically both these places may represent a cult more ancient than that at Annu,[1] since it would appear that in these cases *α* Lyræ was the star personified by Anubis, as *α* Ursæ Majoris and *γ* Draconis were subsequently. But if the cult were more ancient the temple foundations were not, the first "length" of Luxor having been built, on this supposition, about 4900 B.C. The last length built by Rameses II. was certainly oriented to *α* Lyræ, by which I mean that if the building date given by Egyptologists is correct, *α* Lyræ rose in the axis prolonged—another instance of the long persistence of a cult, and of the fact that the temples that we see are but shrines restored.

On the assumption that the above view is true for Luxor and that Abydos followed suit, as is suggested by the imperfect orientations, we are led to the conclusion that, taking existing temple foundations, Annu preceded Abydos.[2]

The astronomical results, then, are certainly in harmony with the historical statement, which I take as fundamental, that Annu preceded Memphis and pyramid times.

These times were not only remarkable on account of the building of the great pyramids; there was a vast change in the cult.

I have already pointed out that at Annu we seemed limited to Set as a stellar divinity; so soon as pyramid times are reached, however, this is changed The number of gods is increased, and there is apparently a mixture, as if some

[1] That is, if we take the temple as oriented originally to *α* Ursæ Majoris.

[2] No sun temple is closely associated with the Set temples either at Luxor or Abydos, and one on the Annu model would not be so associated, for a right angle would carry its axis outside the ecliptic limits.

influence had been at work besides that represented by Annu and the pyramid builders.

I have given before the list of the gods of Heliopolis, and have shown that with the exception of Sit none are stellar. But we find in pyramid times the list is increased; only the sun gods Rā, Horus, Osiris, are common to the two. As new divinities we have[1]—

Isis.
Hathor.
Nephthys.
Ptah.
Serk-t.
Sokhit.

Of these the first two and the last two undoubtedly symbolised stars, and there can be no question that the temples of Isis built at the pyramids, Bubastis, Tanis and elsewhere, were built to watch the rising of some of them.

The temple of Saïs, as I have said, had east and west walls, and so had Memphis, according to Lepsius. The form of Isis at Saïs was the goddess Nit, which, according to some authorities, was the precursor of Athene. The temple of Athene at Athens was oriented to the Pleiades.

There is also no question that the goddess Serk-t symbolised Antares.

We find ourselves, then, in the presence of the worship of the sun and stars in the ecliptic constellations in Egypt during pyramid times, and in constellations connected with the Equinox; for if we are right about the Pleiades and Antares, these are the stars which heralded the sunrise at the Vernal and Autumnal Equinox respectively, when the sun was in Taurus and Scorpio.

[1] Maspero, *op. cit.*, p. 64.

Now, associated with the introduction of these new worships in pyramid times was the worship of the bull Apis, this worship preceding the building of pyramids.

APIS. APIS.

Mena is credited by some authors with its introduction,[1] but at any rate Kakau of the second dynasty issued proclamations regarding it,[2] and a statue of Hapi was in the temple of Cheops.[3]

The ground being thus cleared, I now state the working hypothesis to which I have referred above.

1. The first civilisation as yet glimpsed, so far as temple building goes, in *Northern* Egypt, represented by that at Annu or Heliopolis, was a civilisation with a

[1] Maspero, *op. cit.*, p. 44, note. [2] Maspero, *op. cit.*, p. 64. [3] Maspero, *op. cit.*, p. 46.

non-equinoctial solar worship, combined with the cult of a northern star.

2. Memphis (possibly also Saïs, Bubastis, Tanis and other cities with east and west walls) and the great pyramids were built by a new invading race, representing an advance in astronomical thought. The northern stars were worshipped possibly on the meridian, and a star rising in the east was worshipped at each equinox.

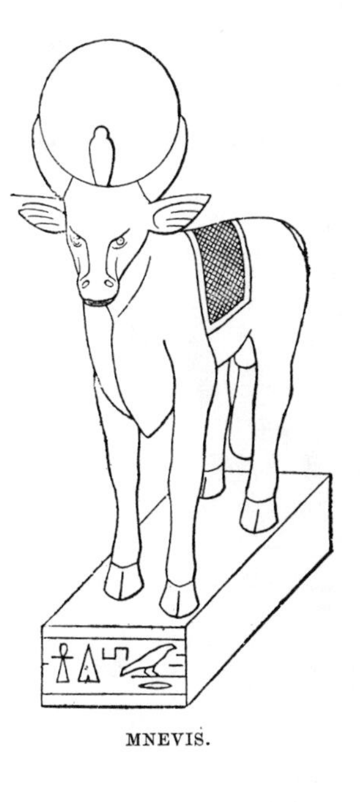

MNEVIS.

3. The subsequent blank in Egyptian history was associated with conflicts between these and other races, which were ended by the victory of the representatives of the old worship of Annu, reinforced from the south, as if north-star and south-star cults had combined against the equinoctial cult.

After these conflicts, east and west pyramid building practically ceased, Memphis takes second place, and Thebes, a southern Annu, so far as the form of solar worship and the cult of Sit are concerned, comes upon the scene as the seat of the twelfth dynasty.[1]

4. The subsequent historical events were largely due to conflicts with intruding races from the north-east. The intruders established themselves in cities with east and west walls, and were on each occasion driven out by solstitial solar worshippers who founded dynasties (eighteenth and twenty-fifth) at Thebes.

[1] Maspero, *op. cit.*, p. 41.

THE TWO GREAT PYRAMIDS AT THE TIME OF THE INUNDATION.

Some detailed remarks are necessary on several points connected with the above generalisation. I will take them *seriatim.*

We find at Memphis, Saïs, Bubastis, and Tanis, east and west walls which at once stamp those cities as differing in origin from Annu, Abydos and Thebes, where, as I have shown, the walls trend either north-west—south-east or north-east—south-west.

For Memphis, Saïs and Tanis the evidence is afforded by the maps of Lepsius. For Bubastis it depends upon the statement of Naville, that the walls run "nearly from east to west," and with the looseness too often associated with such statements, it is not said whether this bearing is true or magnetic.

Associated with these east and west walls there is, moreover, evidence of great antiquity. Bubastis, according to Naville,[1] has afforded traces of the date of Cheops and Chephren, and it is stated by Manetho to have existed as early as the second dynasty.

It is a matter of common knowledge that the pyramids in Egypt are generally oriented east and west.[2] Nor is this all; there has been a distinct evolution in their method of structure.

One of the oldest, if not the oldest pyramid known is the so-called "step-pyramid of Sakkarah.". The steps are six in number, and vary in height from thirty-eight to twenty-nine feet, their width being about six feet. The dimensions are (352 north and south) × (396 east and west) × 197 feet. Some authorities think this pyramid was erected in the first dynasty by the fourth king (Nenephes of Manetho, Ata of the tablet

1 "Bubastis," preface, p. iv.

2 There are, however, notable exceptions to this rule, which will be discussed further on.

THE STEP-PYRAMID OF SAKKARAH.

of Abydos). The arrangement of chambers in this pyramid is quite special.

The claim to the highest antiquity of the step-pyramid is disputed by some in favour of the "false pyramid" of Mêdûm. It also is a genuine step-pyramid, 115 feet high; its outline, which conceals some of the steps, shows three stages, seventy, twenty, and twenty-five feet high; but in its internal structure it is really a step-pyramid of six stages.

THE PYRAMID OF MÊDÛM.

This pyramid must, according to the important and conclusive researches of Professor Flinders Petrie,[1] be attributed to Seneferu, although De Rougé had furnished

[1] "Mêdûm," chap. i.

evidence to the contrary.[1] Seneferu was a king of the fourth dynasty.

We have at Dashûr the only remaining abnormal pyramid, called the blunted pyramid, for the reason that the inclination changes at about one-third of the height. This pyramid forms one of a group of four, two of stone, and, be it carefully borne in mind, two of brick; their dimensions are 700 × 700 × 326 feet; 620 × 620 × 321 feet; 350 × 350 × 90 feet; and 343 × 343 × 156 feet.

THE "BLUNTED PYRAMID" OF DASHÛR.

One of these pyramids was formerly supposed to have been built by Seneferu; if any of them had been erected by King Usertsen III. of the twelfth dynasty, as was formerly thought, the hypothesis we are considering would have been invalid.

Only after Seneferu, then, do we come to the normal Egyptian pyramid, the two largest at Gîzeh built by Cheops

[1] Maspero, *op. cit.*, p. 59.

(Chufu) and Chephren (fourth dynasty) being, so far as is accurately known, the oldest of the series. (According to Mariette the date of Mena is 5004 B.C., and the fourth dynasty commenced in 4235.)

Associated with the cities with east and west walls and these pyramids are temples facing due east, fit, therefore, to receive the rays from a star on the equator or of the morning sun rising at an equinox.

According to Professor Flinders Petrie, at the pyramid of Mêdûm there is a small temple open to the west on the east side of the pyramid. At sunset at the equinox the sepulchral chamber and the sun were in line from the adytum. The priest faced a double Osiris.

Other pyramids were built at Sakkarah during the sixth dynasty, but it is remarkable that such a king as Pepi-Meri-Rā should not have imitated the majestic structures of the fourth dynasty. He is said to have built a pyramid at Sakkarah, but its obscurity is evidence that the pyramid idea was giving way, and it looks as if this dynasty were really on the side of the southern cult, for the authority of Memphis declined, and Abydos was preferred, while abroad Sinai was reconquered, and Ethiopia was kept in order.[1]

The sphinx (oriented true east) may possibly be ascribed to the earliest pyramid builders; it could only have been sculptured by a race with an equinoctial cult.

The Buildings of the Eleventh and Twelfth Dynasties.

We have next to consider what happened after the great gap in Egyptian history between the sixth and twelfth dynasties, 3500 B.C.—2851 B.C. (Mariette); from Nitocris to

[1] Further, it is known that there was some connection between Pepi-Meri-Rā and the eleventh dynasty of Thebes. Maspero, *op. cit.*, p. 91. And it must also be mentioned that in the later pyramids "texts" are introduced.

Amenemhāt I. We pass to the Middle Empire, and here we have merely to deal with the worships previously referred to in Northern Egypt.

Amenemhāt I. built no pyramids, he added no embellishments to Memphis; but he took Annu under his care, and now we first hear of Thebes.[1]

Usertsen I. built no pyramids, he added no embellishments to Memphis, but he also took Annu under his care, and added obelisks to the temples, one of which remains to this day. Further, he restored the temple of Osiris at Abydos, and added to the temple of Amen-Rā at Thebes.[2]

Surely it is very noteworthy that the first thing the kings of the twelfth dynasty did was to look after the only three temples in Egypt of which traces exist, which I have shown to have been oriented to the Sun *not* at an equinox. It is right, however, to remark that there seems to have been a mild recrudescence of pyramid building towards the end of the twelfth dynasty, and immediately preceding the Hyksos period, whether as a precursor of that period or not.

Usertsen's views about his last home have come down to us in a writing by his scribe Mirri:—[3]

> "Mon maître m'envoya en mission pour lui préparer une grande demeure éternelle. Les couloirs et la chambre intérieure étaient en maçonnerie, et renouvelaient les merveilles de construction des dieux. Il y eut en elle des colonnes sculptées, belles comme le ciel, un bassin creusé qui communiquait avec le Nil, des portes, des obélisques, une façade en pierre de Rouou."

There was nothing pyramidal about this idea, but one hundred and fifty years later we find Amenemhāt III. returning both to the gigantic irrigation works and the pyramid building of the earlier dynasties.

[1] Maspero, *op. cit.*, p. 112. [2] Maspero, *op. cit.*, p. 112. [3] Maspero, *op. cit.*, p. 113.

The scene of these labours was the Fayyûm, where, to crown the new work, two ornamental pyramids were built, surmounted by statues, and finally the king himself was buried in a pyramid near the Labyrinth.

The Buildings of the Eighteenth Dynasty.

The blank in Egyptian history between the twelfth and eighteenth dynasties is known to have been associated with the intrusion of the so-called Hyksos. It is supposed these made their way into Egypt from the countries in and to the west of Mesopotamia; it is known that they settled in the cities with east and west walls. They were finally driven out by Aāhmes, the king of solstitial-solar Thebes, who began the eighteenth dynasty.

On page 338 I have shown what happened after the first great break in Egyptian history—a resuscitation of the solar worship at Annu, Abydos and Thebes.

I have next to show that precisely the same thing happened after the Hyksos period (Dyn. 13 (?) Mariette, 2233 Brugsch; Dyn. 18, 1703 B.C., Mariette, 1700 B.C. Brugsch) had disturbed history for some five hundred years.

It is known from the papyrus Sellier (G.C. 257) that Aāhmes, the first king of the eighteenth dynasty, who re-established the independence of Egypt, was in reality fighting the priests of Sutech in favour of the priests of Amen-Rā, the solstitial-solar god, a modern representative of Atmu of Annu.

Amen-Rā was the successor of Menthu. So close was the new worship to the oldest at Annu, that at the highest point of Theban power the third priest of Amen took the same titles as the Grand Priest of Annu, "who was the head of the first priesthood in Egypt." The "Grand Priest of

Annu," who was also called the "Great Observer of Rā and Atmu," had the privilege of entering at all times into the *Habenben* or Naos. The priest Padouamen, whose mummy was found in 1891, bore these among his other titles.

The assumption of the title was not only to associate the Theban priesthood with their northern *confrères*, but surely to proclaim that the old Annu worship was completely restored.

The Buildings of the Twenty-fifth Dynasty.

There was another invasion from Syria, which founded the twenty-second dynasty, and again the government is carried on in cities with east and west walls (Saïs, Tanis and Bubastis). The solstitial-solar priests of Thebes withdraw to Ethiopia. They return, however, in 700 B.C., drive out the Syrian invaders, and, under Shabaka and Taharqa, found a dynasty (the twenty-fifth) at Thebes, embellish the temples there, and at Philæ, Medînet-Habû, and Denderah.

Conclusion.

We see, then, that every important change of cult was associated either with invasions from without or with some disturbance in Egypt itself, for in no other way can the gaps in Egyptian history be explained.

So far we have considered the equinoctial temples as opposed to the non-equinoctial ones in Northern Egypt. We have next to go farther afield, and include the southern temple worship and the possible influence of southern races even in the very earliest times.

CHAPTER XXXIII.

THE CULT OF NORTHERN AS OPPOSED TO SOUTHERN STARS.

So far as my inquiries have yet gone, there is not above Thebes, with the exceptions of Redesieh and Dakkeh, any temple resembling those at Annu, Thebes, Denderah and Abydos, to which I have directed attention as having a high north-east amplitude.

Similarly, with one or two exceptions which are probably late, there are no temples facing the south-east below Thebes.

In short, in Lower Egypt the temples are pointed to rising stars near the north point of the horizon or setting north of west. In Upper Egypt we deal chiefly with temples directed to stars rising in the south-east or setting low in the south-west.

Here again we are in presence of as distinct differences in astronomical thought and purpose of observation as we found among those who directed temples to the sun at the equinox, as opposed to those who worshipped that luminary at some other time of the year.

Now with regard to the northern stars observed rising in high amplitudes, we have found traces of their worship in times so remote that in all probability at Annu and Denderah α Ursæ Majoris was used before it became circumpolar. We deal almost certainly with 5000 B.C.

Since undoubtedly *new* temples with nearly similar amplitudes (such as that denoted by M at Karnak) were built in late times, we find so long a range of time indicated that the utility of the stellar observations *from the yearly point of view* could scarcely have been in question, for the reason that the same star could not herald an equinox or a solstice for four thousand years.

It may be suggested, therefore, that the observations made in them had ultimately to do with the determination of the hours of the night; this seems probable, for in Nubia at present, time at night is thus told.

It may be that such stars as Canopus were used by the southern peoples for the same purpose as α Ursæ Majoris first and then γ Draconis were used by the northerners. In other words, the question arises whether the extreme north and south stars were not both used as warners of the dawn all the year round, after the cult had been established for use at some special time. Canopus, for instance, was of use to herald the autumnal equinox, 6—5000 B.C.; but it is quite natural to suppose that its utility for night work at all times of the year during which it was visible would soon suggest itself, and the same remarks apply to the Northern star γ Draconis.

It is well known that in quite early times means had been found of dividing the day and night into twelve hours. In the day shadows cast by the sun, or sundials, might have been used, but how about the night?

We have seen that the Egyptians chiefly, if not exclusively, observed a heavenly body and the position of other bodies in relation to it, when it was rising or setting, so that it was absolutely essential that the body which they were to observe should rise and set. Everybody knows that as seen in England there are many stars which neither rise nor set. The latitude of London being 51°, the elevation of the pole is 51°.

Hence, any star which lies within that distance from the pole cannot set, but sweeps round without touching the horizon at all. The latitude of Thebes being 25°, the distance from the pole to the horizon is much smaller, and so the number of stars which do not rise and set is much smaller. The stars which do not rise or set are stars near the pole, and therefore stars which move very slowly, and the stars which rise most to the north and most to the south are those bodies which are moving most slowly while they yet rise or set. Can this slow rate of motion have had anything to do with such stars being selected for observation, the brightest star to the north most slowly moving, the brightest star to the south most slowly moving? It is possible that observations of these stars might have been made in such a way that at the beginning of the evening the particular position of γ Draconis, for instance, might have been noted with regard to the pole-star; and seeing that the Egyptians thoroughly knew the length of the night and of the day in the different portions of the year, they could at once—the moment they had the starting-point afforded by the position of this star—practically use the circle of the stars round the north pole as the dial of a sort of celestial clock. May not this really have been the clock with which they have been credited? However long or short the night, the star which was at first above the pole-star,[1] after it had got round so that it was on a level with it, would have gone through a quarter of its revolution.

In low northern latitudes, however, the southern stars would serve better for this purpose, since the circle of northern circumpolar stars would be much restricted. Hence there was a reason in such latitudes for preferring southern

[1] It is worthy of inquiry whether the northern star so observed is not the true Nephthys (Nebt-het). If so, the triad Nephthys, Isis and Horus represents daily astronomical observations.

stars. With regard both to high north and south stars, then, we may in both cases be in presence of observations made to determine the time at night. So that the worship of Set, the determination of the time at night by means of northern stars, might have been little popular with those who at Gebel Barkal and elsewhere in the south had used the southern ones for the same purpose, and this may be one reason why the Theban priests, representing Nubian astronomical culture and methods, were pledged to drive the cult of Sutech out of the land.

Since, then, the observations of γ Draconis might be used to herald the sunrise almost all the year round; and since the modern constellation Draco is the old Hippopotamus, we can readily understand Plutarch's statement that "Taurt presides over the birth of the sun," and why Taurt or Mut should be called the Mistress of Darkness.[1]

It does not seem too much to hope that the continuation of such inquiries may ultimately enable us to solve several points connected with early Egyptian history. We read in Brugsch :—[2]

"According to Greek tradition, the primitive abode of the Egyptian people is to be sought in Ethiopia, and the honour of founding their civilisation should be given to a band of priests from Meroë. Descending the Nile, they are supposed to have settled near the later city of Thebes, and to have established the first state with a theocratic form of government.

"But it is not to Ethiopian priests that the Egyptian Empire owes its origin, its form of government, and its high civilisation; much rather was it the Egyptians themselves

[1] Rawlinson, i. 337.

[2] "Egypt under the Pharaohs," ed. 1891, p. 3.

that first ascended the river to found in Ethiopia temples, cities, and fortified places, and to diffuse the blessings of a civilised state among the rude dark-coloured population.

. . . "Strange to say, the whole number of the buildings in stone, as yet known and examined, which were erected on both sides of the river by Egyptian and Ethiopian kings, furnish incontrovertible proof that the long series of temples, cities, sepulchres, and monuments in general, exhibit a distinct chronological order, of which the starting-point is found in the pyramids, at the apex of the Delta."

It must be emphatically stated that the results obtained from these monuments, studying them from the astronomical point of view, lead to a very different conclusion. Instead of one series, there are distinctly two (leaving out of consideration the great pyramid builders at Gîzeh) absolutely dissimilar astronomically; and instead of one set of temple-builders going up the river, there were at least two sets, one going *up* the river building temples to north stars, the other going *down* building temples to south stars; and the two streams practically met at Thebes, or at all events they were both very fully represented there, either together or successively.

The double origin of the people thus suggested on astronomical grounds may be the reason of the name of "double country," used especially in the titles of kings, of the employment of two crowns, and finally of the supposed sovereignty of Set over the north, and of Horus over the south divisions of the kingdom.[1]

Only by the time of Seneferu was there anything like an amalgamation of the peoples. He first was "King of the two Egypts,"[2] while later Chephren called himself "Horus and Sit"[3]

[1] Brugsch, "History," p. 6. [2] Maspero, "Histoire ancienne," p. 59. [3] *Idem*, p. 63.

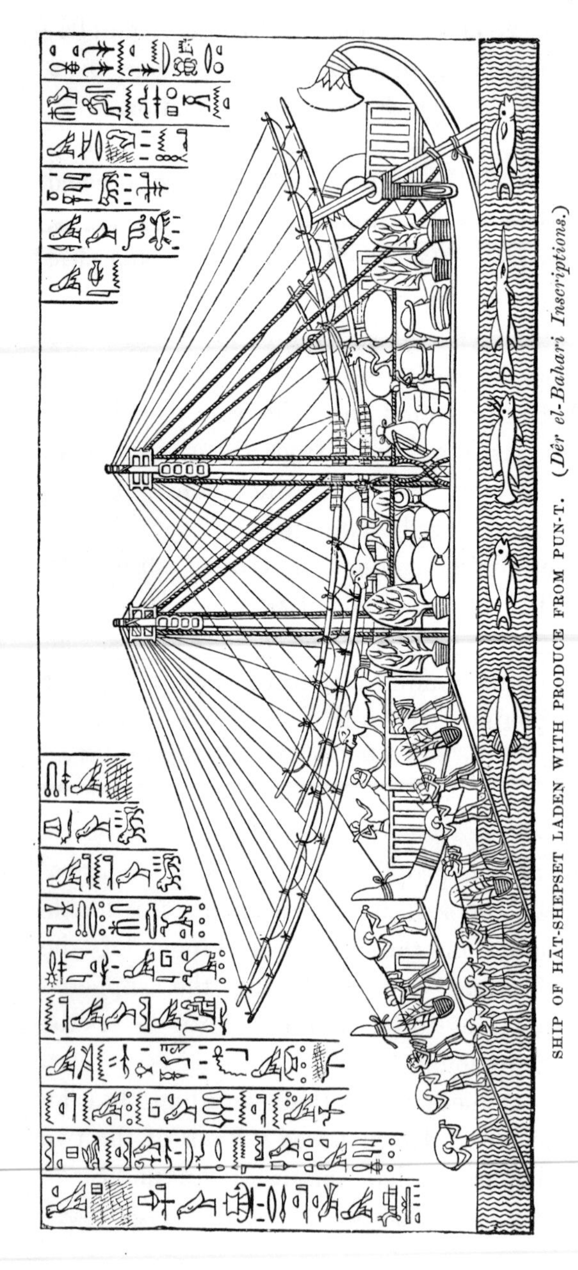

SHIP OF HĀT-SHEPSET LADEN WITH PRODUCE FROM PUN-T. (*Dêr el-Bahari Inscriptions.*)

—a distinct indication, I take it, that the influence of Upper Egypt was already felt as early as Seneferu, and, I think, much earlier, although all *temple* trace of it is lost.

With regard to the start-point of the temple-builders who came down the river, there is no orientation evidence, for the reason that there is little or no information from the regions south of Naga. At Naga (lat. 16° 18′ N.), Meroë (lat. 16° 55′ N.), Gebel Barkal and Nuri (both in lat. 18° 30′ N.), there is information of the most important kind, but beyond Naga there is a gap; but since important structures were erected at the places named in early times (my inquiries suggest 3000–4000 B.C.), it is probable that the peoples who built them stretched further towards the equator.

But although the orientation evidence is lacking for the lower latitudes, the inscriptions are by no means silent, and over and over again it is stated that those particular

HUTS BUILT ON PILES IN PUN-T. (*Dêr el-Bahari Inscriptions.*)

gods whom I have found to be associated with southern stars came from a locality called the land of Pun-t.

Pun-t was always considered a "Holy Land." Hathor was "Queen of the Holy Land," "Mistress and Ruler of Pun-t." Amen-Rā was "Hak" or "King" of Pun-t, and Horus was the Holy Morning Star which rose to the west (?) of the land of Pun-t.[1]

Maspero refers to an ancient tradition that the land of Pun-t could be reached by going up the Nile, where eventually one came to an unknown sea which bathed the land of Pun-t. Was this one of the great lakes?[2]

[1] Rawlinson, ii., p. 134.

[2] Maspero, "Histoire ancienne," p. 5.

Brugsch[1] is of opinion that Pun-t occupied the south and west coasts of Arabia Felix, but Maspero and Mariette do not agree with him. The two latter authorities identify it with that part of the Somali-land which borders on the Gulf of Aden. It is the Cinnamonifera regio or Aromatifera regio of the ancients.[2]

The inscriptions at Dêr el-Bahari make it quite certain that Pun-t is in Africa. Hottentot Venuses, pile dwellings, elephants, to say nothing of the products of the country referred to as among the freight of the ships on their homeward voyage, distinctly point to Africa, and I think a southern part of it. The Cynocephalus ape, perhaps, is more doubtful.

The first organised expedition to Pun-t of which we hear anything is that organised by Se-ānχ-ka-Rā, the last king of the 11th *Theban* dynasty. This was a new traffic by way of the Red Sea. There was then no canal in existence

[1] Brugsch, "History of Egypt," 1891, p. 54.

[2] Mariette, "Dêr el-Bahari," p. 31. Mr. W. T. Thistleton-Dyer, the director of the Royal Gardens, Kew, agrees in this view. He permits me to print the following extract from a letter written to me:—"The only positive fact that I can deal with is the representation in the pictures of a small scrubby tree, which seems to have been about four feet high. It appears to have yielded a gummy or resinous exudation from its trunk. Mariette supposes this to be myrrh, Pount to be Somali-land, and To Nuter the Socotran Archipelago. All this fits in very well with botanical facts. Myrrh-producing plants exist both in Somali-land and Arabia, and also in Socotra, as ascertained by Bayley Balfour. The two former places still are, as they always have been, the place of origin of myrrh, and we know that it was largely used by the Egyptians in embalming. There is no evidence that myrrh, or anything in any way resembling it, was ever found south of the Equator. I cannot carry you further south than Berbera."

[3] On this point I am permitted to print the following extract from a letter received from my friend Sir John Kirk, K.C.B.:—"I send you a photo, taken in 1858, in the delta of the Zambezi, of a house built on high poles. The people there live in such houses. There is a ladder by which they mount, and all their belongings are kept above. Such houses I have since seen at the mouth of the River Rufiji, opposite the island of Monfia, to the south of Zanzibar. The reason in both cases for such a type of house is that the country at one time is flooded, and also to avoid mosquitoes. Similar structures are used, I am told, in Madagascar. At Lake Nyassa I believe there are village communities living in the lake, on artificial islands of piles."

joining the sea with the Nile; the expedition went by land to Coptos.[1]

They further indicate, as Maspero suggests, that the expedition of Hātshepset anchored up a river, and not on the sea-shore. This, again, makes Africa much more probable than Arabia.

CYNOCEPHALUS APE WITH MOON EMBLEM.

If we agree that Pun-t is really in Africa, south of Somali-land, there is a great probability that the tradition referred to by Maspero is a true one.

It is also to be pointed out that there is no trace of the southern star temples along the various roads to the Red Sea, while, on the other hand, the earliest traces of northern star worship, with the exception of Annu or On, occur along one or other of them. There is distinct evidence that Osiris, Horus, Hathor, Chnemu, Amen-Rā, and Khons, are worships coming from the south. With regard to Horus, it is necessary to discriminate, since there were two distinct gods—Horus in Northern and Horus in Southern Egypt, *and Horus of the south was the elder of the two.*

The Hawk-god of Edfû, Harhouditi, the southern Horus, had for servants a number of individuals called Masniu or Masnitiu=blacksmiths. The Hawk-god of the Delta, the northern Horus, Harsiisit, had for his entourage the Shesu Horu.

Now Maspero has recently pointed out[2] that the southern Horus may have been imported, not from Arabia Felix or Somali-land, but from Central Africa! and in a most interesting paper has called attention to some customs still extant among

[1] Rawlinson, ii., p. 131. [2] "L'Anthropologie," 1891, No. 4.

the castes of blacksmiths in Central Africa, which have suggested to him that the followers of the Edfû Horus may have come from that province.

He writes:—

"C'est du sud de l'Égypte que les forgerons sont remontés vers le nord; leur siège primitif était le sud de l'Égypte, la partie du pays qui a le plus des rapports avec les régions centrales de l'Afrique et leurs habitants."

Then, after stating the present conditions of these workers in Equatorial Africa, where they enjoy a high distinction, he concludes:—

"Je pense qu'on peut se représenter l'Horus d'Edfou comme étant au début, dans l'une de ses formes, le chef et le dieu d'une tribu d'ouvriers travaillant le métal, ou plutôt travaillant le fer. On ne saurait en effet se dissimuler qu'il y a une affinité réelle entre le fer et la personne d'Horus en certains mythes. Horus est la face céleste (horou), le ciel, le firmament, et ce firmament est de toute antiquité, un toit de fer, si bien que le fer en prit le nom de ba-ni-pit, métal du ciel, métal dont est formé le ciel: Horus l'aîné, Horus d'Edfou, est donc en réalité un dieu de fer. Il est, de plus, muni de la pique ou de la javeline à point de fer, et les dieux qui lui sont apparentés, Anhouri, Shou, sont de piquiers comme lui, au contraire des dieux du nord de l'Égypte, Rā, Phtah, etc., qui n'ont pas d'armes à l'ordinaire. La légende d'Harhouditi conquérant l'Égypte avec les masniou serait-elle donc l'écho lointain d'un fait qui se serait passé au temps antérieurs à l'histoire? Quelque chose comme l'arrivée des Espagnols au milieu des populations du Nouveau Monde, l'irruption en Égypte de tribus connaissant et employant le fer, ayant parmi elles une caste de forgerons et apportant le culte d'un dieu belliqueux qui aurait été un Horus ou se serait confondu avec l'Horus des premiers Egyptiens pour former Harhouditi. Ces tribus auraient été nécessairement d'origine Africaine, et auraient apporté de nouveaux éléments Africains à ceux que renfermait déjà la civilisation du bas Nil. Les forgerons auraient perdu peu à peu leurs privilèges pour se fondre au reste de la population: à Edfou seulement et dans les villes où l'on pratiquait le culte de l'Horus d'Edfou, ils auraient conservé un caractère sacré et se seraient transformés en un sorte de domesticité religieuse, les masniou du mythe d'Horus, compagnons et serviteurs du dieu guerrier."

If we are to accept Maspero's suggestion that the elder Horus really came from Central Africa, traces of the cult of his followers should be found high up the river.

But such a search is now denied us, while in the time of Thothmes III. it is supposed that the south frontier Kali of the inscriptions is probably connected with Koloë in 4° 15′ N. lat. according to Ptolemy.[1]

As a matter of fact, there is distinct evidence of the cult of the southern stars coming down the river in the region we can get at; α Centauri, *e.g.*, seems to have been observed at Gebel Barkal before Thebes—Sirius is too modern to be considered—and above all there is the remarkable series of temples, apparently oriented to Canopus before 6000 B.C., which come down no lower than Edfû.

The general statement is, then, that there were two distinct groups of stellar temples, probably built by different races, or at all events by peoples having very different astronomical methods.

It is well to inquire here whether the dates of the various temples as determined by the methods dwelt on in previous chapters can throw any light upon the inquiry. Here I must re-state that in almost every case the date of foundation so determined precedes the generally-received date, which invariably has reference to a stone building, while in all probability the first structure was a brick shrine merely. In support of this view I may state that the looking after ruined shrines was recognised as one of the duties of kingship.

> "I have caused monuments to be raised to the gods; I have embellished their sanctuaries that they may last to posterity; I have kept up their temples; I have restored again what was fallen down, and have taken care of that which was erected in former times." [2]

Not only did Thothmes III. find the original temple of Amen-Rā built in brick, but he found the temple at Semneh in brick also, and he rebuilt it in memory of Usertsen III.[3]

[1] Brugsch, "Egypt," p. 184.
[2] Inscription of Thothmes III., translated by Brugsch, "Egypt," p. 188.
[3] Brugsch, "Egypt," p. 184.

In the following table I bring together the foundation dates I have found most probable, bearing the above and many other considerations in mind. The dates are, of course, only provisional, since local data are in many cases wanting. Where no information is forthcoming as to the height of the horizon visible along the temple axis, I have assumed hills 1° high, and used the dates printed in heavy type in Chapter XXX.

TABLE OF TEMPLES BUILT TO N. AND S. STARS.

Years B.C.	Northern Stars.				Southern stars.				Remarks.
	α Ursæ Majoris.	γ Draconis.	Capella.	Spica.	Phact.	α Centauri.	Canopus.	Sirius.	
[6400]							1, 2, 3, 4,		1 2 3 4 ? Edfû, Philæ, Amada, Semneh.
5400									
5300			1						1. Memphis.
			2						2. Annu.
5200	1								1. Annu.
5100									
5000									
4900									
4800	1								1. Denderah.
4700									
4600		1							1. Redesieh.
4500									
4400									
4300									
4200	1	2							1, 2. Denderah (temple built when both stars had an equal amplitude).
4100									
4000									
3900									
3800									
3700						1			1. Barkal (E).
						2			2. Kûrnah (Seti I.).
					3				3. Memnonia (Western Temple).
3600						1			1. Kûrnah (Palace).
					2				2. Barkal (B).
3500									1. Karnak (Z and X)
		2							2. Dakkeh.
		3							3. Denderah.
3400					1				1. Karnak (V).
3300									

TABLE OF TEMPLES BUILT TO N. AND S. STARS (*continued*).

Years B.C.	Northern Stars.				Southern Stars.				Remarks.
	α Ursæ Majoris.	γ Draconis.	Capella.	Spica.	Phact.	α Centauri.	Canopus.	Sirius.	
3200					1				1. Abu Simbel (Hathor Temple).
				2					2. Karnak (Y).
3100								1	1. Karnak (Temple O) Gr.
					2				2. Dêr el-Medînet (Gr.).
3000			1						1. Karnak (U).
						2			2. Wady Halfa (Thothmes II.).
2900						1			1. Barkal (L).
2800						1			1. Wady Halfa (Thothmes III.).
					2				2. Saboоa.
2700								1	1. Dêr el-Bahari.
						2			2. Wady E. Sofra.
2600						1			1. Memnonia (Rameses II.) (Mean of Fr. and Gr.).
2500		1							1. Karnak (W).
					2				2. Karnak (J).
					3				3. Medînet Habû (JJ).
2400						1			1. Kom Ombo (Little Temple).
			2						2. Petit Temple du Sud (Memnonia).
					3				3. Barkal (J and H).
2300		1							1. Annu (Restoration).
2200									
2100							1		1. Karnak (B).
					2				2. Semneh.
2000								1	1. Dosche.
1900				1					1. Tell el-Amarna.
1800								1	1. Karnak (D) Gr.
1700			1						1. Karnak (G).
							2		2. Karnak (Seti II.).
1600									
1500									
1400								1	1. Naga (Temple g), Gr.
1300							1		1. Naga (Temple f).
1200		1							1. Karnak, (A.M.C).
				2					2. Medînet Habû (Palace K K).
1100									
1000							1		1. Karnak (Khons).
900									
800								1	1. Philæ (Ethiopian Temple).
					2				2. Medînet Habû (Ethiopian Temple).
700								1	1. Denderah (Isis Temple).

Gr. = German values of Orientation.
Fr. = French ”

The following general conclusions may be drawn from the table :—

I. At the earlier periods there are well-marked epochs of temple-building revealed by the table.

II. If we can accept the possible Canopus temples referred to in Chapter XXX., the oldest foundations in Egypt yet traced are to southern stars. They are limited to Upper Egypt, and date from before 6000 B.C.

III. The temples to the north stars, *α* Ursæ Majoris, *γ* Draconis, and Capella (Set and Ptah), begin in the Delta and about 1000 years later. The series is then broken till about 3500 B.C.

IV. The south star temples to Phact at the summer solstice, and *α* Centauri at the autumnal equinox, begin about 3700 B.C.

V. *γ* Draconis replaces *α* Ursæ Majoris at Denderah; north-star temples are for the first time erected in the south at Karnak and Dakkeh in 3500 B.C.

VI. For the first time about 3200 B.C., north- and south-star temples are built simultaneously.

VII. After this the building activity is chiefly limited to temples to southern stars.

If we take Brugsch's dates, we find that the foundations of the greatest number of temples were laid about the time of Seneferu, Pepi, and the twelfth dynasty. The more modern kings founded few temples—their functions were those of expanding, restoring, and *annexing*. Even Thothmes III. seems to have laid no new foundations except perhaps that of the Ptah temple at Karnak, and that is doubtful.

The wonderful Hall of Columns called Khu-mennu (Splendid Memorial), in the temple of Amen-Rā, was dedicated by Thothmes III. not only to Amen-Rā, but to his ancestors. It is important to see who these were in the present connection. I give them with approximate dates.[1]

	Brugsch. B.C.	Mariette. B.C.
Seneferu, fourth dynasty	3766 ...	4235
Assa, fifth dynasty	3366 ...	3951
Pepi, sixth dynasty	3233 ...	3703
The Antefs, eleventh dynasty	2500 ...	3064
The most famous sovereigns of the twelfth dynasty	2433—2300 ...	2851
Thirty princes of the thirteenth dynasty	2233	(?)

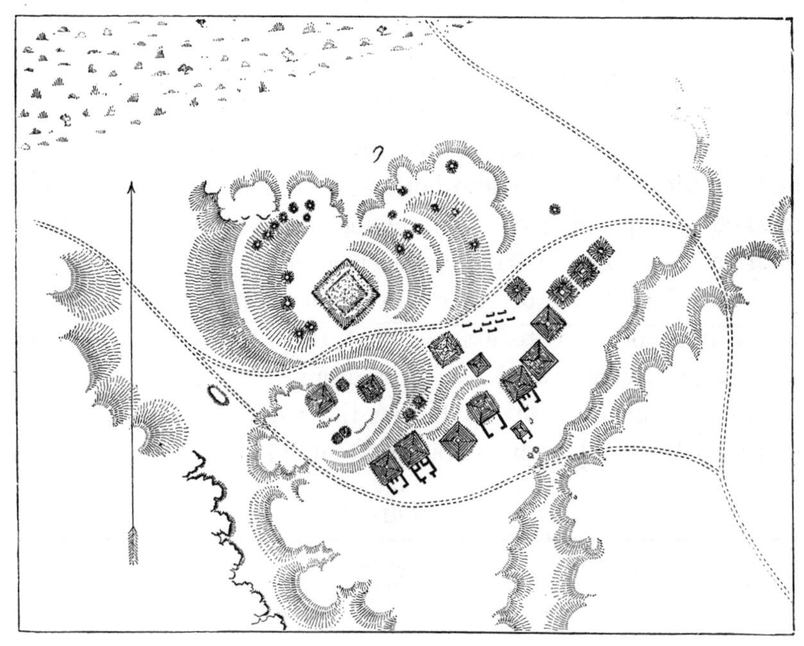

PLAN OF THE PYRAMIDS AT NURI.

It is interesting to note that in this list the builders of the great pyramids at Gîzeh, and all the kings who in the last chapter were suggested as being given to equinoctial worship, are passed over without notice. It would appear,

[1] Brugsch, "Egypt," p. 180.

then, that the ancestors named were of southern origin, precursors of Thothmes in cult as well as in time.

Of these ancestors, the first—if Brugsch's dates can be taken, which, I think, is doubtful—limited himself to southern temples; the majority of temples built near Pepi's time were oriented to the south. The twelfth dynasty was more catholic.

The more we inquire, the more interesting does this inquiry into the north-star temples as opposed to the south-star temples become. These considerations are not limited to the temples—they apply also to pyramids. At Gîzeh we find both temples and pyramids oriented east and west. At Gebel Barkal, Nuri, and Meroë, in Upper Egypt, we find both temples and pyramids facing south-east, and at the first of these places, where both exist together, we find well-marked groups of pyramids connected by their orientations with each temple. I can, however, find no information as to the probable dates of these pyramids; in the absence of facts, it seems fair to assume that they follow the dates of the temples which agree in orientation.

In the following tables I give the values for Nuri, Meroë, and Gebel Barkal; a west variation of $8\frac{1}{2}°$ has been assumed.

NURI.[1]

Cult.	Magnetic Azimuth.	Astronomical Amplitude.	Decl.
	°	°	°
Pyramids 10, 11, 12 ...	N. 136 E.	$37\frac{1}{2}$ S. of E.	S. $35\frac{1}{4}$
Pyramids 1, 4	N. $137\frac{1}{4}$ E.	$38\frac{3}{4}$ S. of E.	S. $36\frac{1}{4}$
Pyramids 13, 14, 15 ...	N. 139 E.	$40\frac{1}{2}$ S. of E.	S. 38
Pyramids 2, 3, 16, 17 ...	N. $145\frac{1}{2}$ E.	47 S. of E.	S. $43\frac{3}{4}$
Pyramids 5, 6, 7, 8, 9 ...	N. $146\frac{1}{2}$ E.	48 S. of E.	S. $44\frac{3}{4}$

[1] For plans, *see* Lepsius, vol. ii., p. 130.

MEROË.[1]

Cult.	Magnetic Azimuth.	Astronomical Amplitude.	Decl.
	°	°	°
Pyramid 16	N. 102 E.	3½ S. of E.	S. 3¼
Pyramid 20	N. 103 E.	4½ S. of E.	S. 4¼
Temple near Watercourse ...	N. 112 E.	13½ S. of E.	S. 12¾
Pyramid 15	N. 112 E.	13½ S. of E.	S. 12¾
Pyramids 14, 37	N. 113 E.	14½ S. of E.	S. 13¾
Pyramid 10	N. 116 E.	17½ S. of E.	S. 16¾
Pyramid 39	N. 118 E.	19½ S. of E.	S. 18¾
Pyramid 19	N. 83 E.	15½ N. of E.	N. 14¾

GEBEL BARKAL.[2]

Cult.	Magnetic Azimuth.	Astronomical Amplitude.	Decl.
	°	°	°
Temple E	N. 132 E.	33½ S. of E.	S. 31½
Pyramid 18	N. 132½ E.	34 S. of E.	S. 32
Temple L	N. 136½ E.	38 S. of E.	S. 35½
Pyramids 9, 13	N. 136 E.	37½ S. of E.	S. 35¼
Pyramid 11	N. 140 E.	41½ S. of E.	S. 39
Pyramids 1, 2	N. 141 E.	40½ S. of E.	S. 39¾
Temples J and H	N. 146 E.	47½ S. of E.	S. 44¼
Pyramid 20	N. 146 E.	47½ S. of E.	S. 44¼
Pyramids 2, 15, 16, 17 ...	N. 147 E.	48½ S. of E.	S. 45¼
Temple B	N. 152 E.	53½ S. of E.	S. 49¾
Pyramids 5, 6, 7, 8, 10 ...	N. 153 E.	54½ S. of E.	S. 50½
Pyramid 19	N. 156 E.	57½ S. of E.	S. 53

[1] For plans, *see* Lepsius, vol. ii., pp. 133 and 134.
[2] For plans, *see* Lepsius, vol. ii., pp. 125 and 127.

It seems quite justifiable from the above facts to conclude that the pyramids and temples oriented S.E. and, as I hold, to *α* Centauri when it heralded the autumnal equinox, were not built by people having the same astronomical ideas, worships, and mythology as those who built at Gîzeh due E. and W., and marked the autumnal equinox by the heliacal rising of Antares.[1] The only thing in common was noting

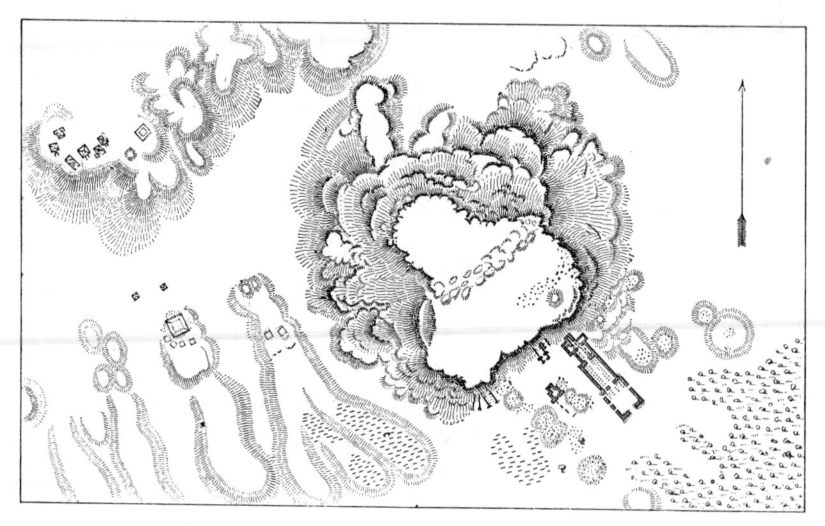

PLAN OF THE TEMPLES AND PYRAMIDS AT GEBEL BARKAL.

an equinox, and so far as this goes we may infer that neither people dwelt originally in the Nile Valley, but came by devious ways from a country or countries where the equinoxes had been made out.

[1] There is a point of great interest here. It would seem from Captain Lyons' examination of the temples at Wady Halfa, which I make out to have been oriented to *α* Centauri, that when the two races were amalgamated in later times, both the stars to which I have referred as heralding the equinox were personified by the same goddess, Serk-t.

CHAPTER XXXIV.

THE ORIGIN OF EGYPTIAN ASTRONOMY—THE NORTHERN SCHOOLS.

So far we have dealt with the dawn of astronomy in Egypt. We have found that from the earliest times there were astronomical observations carried on, and that practically there were three schools of thought. To all three schools sun-worship was common, but we may clearly separate them by the associated star-worship. We have found worshippers of northern stars, east and west stars, and southern stars.

The northern star-worshippers we may associate with Annu, the east and west star cult with the pyramid fields at Gîzeh, and the southern star-worshippers with Upper Egypt.

What we have to do in the present chapter is to see whether the orientation of the structures helps us with any suggestions touching the question whether we have to stop at the places named and acknowledge Egypt to be the true cradle of astronomical science; or whether the facts we have considered compel us to go a stage further back, and to recognise that the true origin was elsewhere; that, in short, astronomy, instead of taking its rise in Egypt, was simply imported thither.

It would appear from the recent work of the students of the languages of Babylonia and Assyria that in these countries, if anywhere, there might have been civilisations more ancient than the Egyptian, which have already been glimpsed.

But before I go further something must be said about Babylonia itself, for the reason that it also was the meeting-ground of at least two different schools of astronomical thought.

The facts connected with this subject are still to a certain extent involved in obscurity, which is little to be wondered at when we think how recently any knowledge has been available to throw light upon the past of these regions. I need, however, only briefly refer to them, and for this purpose shall use the two most recently published books dealing with the question which at present concerns us. I refer to Prof. Sayce's "Hibbert Lectures" and Prof. Jensen's "Kosmologie der Babylonier."

But what period are we to take?

It follows from the investigation into the orientation of Egyptian temples that the stars *a* Ursæ Majoris, Capella, Antares, Phact, and *a* Centauri were carefully observed, some of them as early as 5000 B.C., the others between 4000 and 3000 B.C. I have also shown that it is possible that at Edfû and Philæ the star Canopus may have been observed as early as 6400 B.C. Further, that the constellations of the Thigh (Ursa Major), the Hippopotamus (Draco), the Bull, and the Scorpion had been established in pyramid times.

It becomes important, therefore, if we recognise this as the dawn of astronomy in Egypt, to see if any information is extant giving us information concerning Babylonia, so that we may be able to compare the observations made in the two regions, not only with a view of tracing the relative times at which they were made, but to gather from these any conclusions that may be suggested in the course of the inquiry.

The inquiry must be limited to certain detailed points; we know quite well already, as I have stated before, that the omen tablets, which mention a king called Sargon (probably Sargon I. of Agade), who reigned in Babylon about 3700 B.C., prove unquestionably that astronomy had been cultivated for thousands of years before the Christian

Era.[1] But to institute a comparison we must leave the general and come to the particular. I will begin with the northern constellations, as it follows from my researches that very early at Annu and Denderah temples were erected for their worship—the worship of Anubis or Set, as I have shown before; that is, of α Ursæ Majoris and γ Draconis.

THE ANNU SCHOOL. THE WORSHIP OF SET.

According to Maspero, Set formed one of the divine dynasties at Annu, and the northern stars seem to have been worshipped there. I suppose there is now no question among Egyptologists that the gods Set, Sit, Typhon, Bes, Sutekh, are identical. To this list possibly Ombo and Nubi should be added.[2] It is also equally well known that Sutekh was a god of the Canaanites,[3] and Bes is identified with Set in the Book of the Dead.[4]

It is also stated by Maspero that at Memphis[5] (time not given) there were temples dedicated to "Sutekh" and "Baal." In the chapter on the circumpolar stars I have suggested that they were taken as typifying the powers of darkness and of the lower world, and I believe it is conceded by Egyptologists that Anubis in jackal form was either contemporaneous with or preceded Osiris in this capacity.

In the exact centre of the circular zodiac of Denderah we find the jackal located at the pole of the equator; it obviously represents the present Little Bear.

[1] Besides the book on omens we have "The observations of Bel," or "Illumination of Bel" (Mul-lil), seventy-two books dealing with conjunctions of Sun and Moon, phases (?) of Venus, and appearance of comets. (Sayce, "Hibbert Lectures," p. 29.) The complete materials for the study of Babylonian and Assyrian astronomy cannot be available until the catalogue of the Kouyunjik Collection, now in course of publication by the British Museum, is finished.

[2] Rawlinson, vol. i., p. 316.

[3] Maspero, "Histoire Ancienne," p. 165.

[4] Pierret, "Le Panthéon Égyptien," p. 48. I have before referred to a doubt on this point.

[5] Maspero, *op. cit.*, p. 357.

Now, do we get any Babylonian connection so far as we have gone?

We learn, to begin with, from Pierret[1] that the hippopotamus, the emblem of Set and Typhon, was the hieroglyph of the Babylonian god "Baal."

Do we get the jackal constellation in Babylonian astronomy? Of this there is no question, and in early times. Jensen refers[2] to the various readings "jackal" and "leopard," and states that it is only doubtful whether by this figure the *god* ANU or the *pole of the ecliptic* ANU is meant. Either will certainly serve our present purpose, and a leopard in Babylonia might as easily symbolise the night as a jackal in Northern Egypt.

There seems little doubt that the jackal, leopard, hyæna, black pig (wild boar), and hippopotamus were chosen as the representatives of the god of evil and darkness (associated with the circumpolar constellations), on account of their ravages on flocks and herds and crops. If this be agreed, nothing is more proper than that the jackal should be associated with North Egypt, the hippopotamus with South Egypt, and the wild boar with a latitude to the north of Egypt (and perhaps of Nineveh) altogether. The representative of the god of darkness, then, depended upon the latitude. In this connection I may state that Drs. Sclater and Salvin have quite recently referred me to an interesting paper by the late Mr. Tomes[3] on the habit of the hippopotamus when it comes out of the water to exude a blood-coloured fluid from special pores in its skin. This explains at once why Typhon took the form of a *red* hippopotamus, and why

[1] "Le Panthéon Égyptien," p. 4.
[2] "Kosmologie der Babylonier," p. 147.
[3] "Proc. Zool. Soc.," 1850, p. 160.

Mr. Irving, on the modern stage, couples Mephistopheles, the modern devil, with red fire.

I know not whether the similarity in the words Anu, Annu and An results merely from a coincidence, but it is certainly singular that the most ancient temples in Lower Egypt (Heliopolis and Denderah) should be called Annu or An[1] if there be no connection with the Babylonian god Anu.

With regard to Anubis, it is quite certain that the seven stars in Ursa Minor make a very good jackal with pendent tail, as generally represented by the Egyptians (*see* page 276), and that they form the nearest compact constellation to the pole of the ecliptic.

The worship of Anubis as god of the dead, or the night god, whether associated with the Babylonian Anu or not, was supreme till the time of Men-Kau-Rā, the builder of the third pyramid of Gîzeh[2] (3633 B.C., Brugsch; 4100 B.C., Mariette). Osiris is not mentioned. The coffin-lid of this king with the prayer to Osiris "marks a new religious development in the annals of Egypt. The absorption of the justified soul in Osiris, the cardinal doctrine of the Ritual of the Dead, makes its appearance here for the first time."

It seems extremely probable, therefore, that the worship of the circumpolar stars went on in Babylonia as well as in Egypt in the earliest times we can get at.

A very wonderful thing it is that, apparently in very early times, the Babylonians had made out the pole of the equator as contradistinguished from the pole of the ecliptic. This they called Bīl. With this Jensen finds no star associated,[3] but 6000 B.C. this pole would be not far removed from those

1 Dr. Wallis Budge informs me that An was an old name of the sun-god.
2 Rawlinson, vol. ii., p. 64.
3 "Kosmologie der Babylonier," p. 147.

stars in the present constellation Draco, out of which I have suggested that the old Egyptian asterism of the hippopotamus was formed.

Nor was this all; movements in relation to the ecliptic had been differentiated from movements in relation to the equator. We have inscriptions running:—

"*The way in reference to Anu,*" that is the ecliptic with its pole at Anu.
"*The way in reference to Bīl,*" the equator with its pole at Bīl.

In other words, the daily and yearly apparent movements of the heavenly bodies were clearly distinguished, while we note also

Kabal šami, "the middle of the heavens," defining the meridian.

So far as I can make out, when Anubis was supreme in Egypt, the only sun-gods at Memphis and Annu were Rā and Atmu. Ptah appears to have been a mixed sun-star god, *i.e.*, Capella heralding the sun-rise in the Harvest Time.

Now I learn from Prof. Sayce[1] that in Babylonia Anu and Bīl ranked as two members of a triad from the commencement of the Semitic period, the third member being probably a southern star symbolised as we shall see in the sequel; it is only in later times in Babylonia apparently that we get a triad consisting of sun, moon, and Venus,[2] Venus being replaced at Babylon by Sirius.[3]

To the two northern divinities temples were built; both were worshipped in one temple at Babylon,[4] which must therefore have been oriented due north; and the pole of the equator (the altitude of which is equal to the latitude of the place) was probably in some way indicated. Here there was no rising or setting observation, for Eridu, the most southern of the old Babylonian cities, had about the same latitude as Bubastis,

[1] Sayce, "Hibbert Lectures," 1887, p. 193.
[2] Sayce, p. 193.
[3] Jensen, p. 149.
[4] Sayce, p. 439.

in Egypt. The pole of the ecliptic (Anu) would revolve round the pole of the equator (Bīl) always above the horizon.

So that since Sutekh = Anu
and Baal = Bīl,

the temple at Memphis to those divinities reported by Maspero (see *ante*) must have been oriented in the same way as the one at Babylon, that is to the north; and if the above evidence be considered strong enough to enable us to associate the Babylonian Bīl with the Egyptian Taurt, we have not only Ursa Minor but Draco represented in the early worship and mythology both of Egypt and of Babylonia.

According to Prof. Sayce[1] there is distinct evidence of a change of thought with regard to Anu in Babylonia—there certainly were great changes of thought in Egypt with regard to Anubis. Observations of stars near the pole of the ecliptic appear to have been utilised before they were taken as representing either the superior or inferior powers—before, in fact, the Anubis or Set stage *quâ* Egypt was reached. After this had been accomplished there was still another advance, in which Anu assigns places to sun, moon, and evening star, and symbolises the forces of nature.

There is evidence, though unfortunately it is very meagre, that the temple worship was very similar in the two countries.

In the ceremonials in the temples the statues of the gods in boats or arks were always carried in procession.[2] The same rectangular arrangement of temples which held in Egypt, held also in Babylonia, and this perhaps may be the reason why Bīl seems so often to refer to the sun, whereas it was the name given to the combined worship. Sometimes, on the other hand, the worship of the stars is distinctly referred to as taking

[1] Sayce, "Hibbert Lectures," p. 190.

[2] Sayce, p. 280. There is a bas-relief in the British Museum showing this ceremonial.

place in a solar temple. Thus at Marduk's temple, E-Sagili, we are told that "two hours after nightfall the priest must come and take of the waters of the river; must enter into the presence of Bīl, and putting on a stole in the presence of Bīl must say this prayer," etc.[1] The temple, then, will probably have been oriented to the north. Night prayers in a sun-temple afford pretty good indications of a mixed cult.

The evidence, then, seems conclusive that by the time of the founding of the temple at Annu a knowledge of the stars near the pole of the equator, and of the importance of observing them, was common to N. Egypt and to the region N.E. of it. Whether the worship of Set was introduced into Egypt from this region, or whether there was a common origin, must for the present, then, remain undetermined.

THE EQUINOCTIAL SCHOOL—THE WORSHIP OF THE SPRING-SUN.

The East and West orientation, as we have seen, is chiefly remarkable at the pyramids of Gîzeh and the associated temples, but it is not confined to them.

The argument in favour of these structures being the work of intruders is that a perfectly new astronomical idea comes in, one not represented at Annu and quite out of place in Egypt, with the solstitial rising river, as the autumnal equinox was at Eridu, with the river rising at the spring equinox.

We are justified from what is known regarding the rise of the Nile as dominating and defining the commencement of the Egyptian year at the solstice, in concluding that other ancient peoples placed under like conditions would act in the same way; and if these conditions were such that spring would mean sowing-time and autumn harvest-time, their year would begin at an equinox.

[1] Sayce, p. 101.

Now what the valley of the Nile was to Egypt those of the Tigris and the Euphrates were to the Babylonian empire. Like the Nile, these valleys were subject to annual inundations, and their fertility depended, as in Egypt, upon the manner in which the irrigation was looked after.

But unlike the Nile, the commencement of the inundation of these rivers took place near the vernal equinox; hence the year, we may assume, began then, and, reasoning by analogy, the worship in all probability was equinoctial.

A people entering Egypt from this region, then, would satisfy one condition of the problem. But is there any evidence that this people built their solar temples and temple walls east and west, and that they also built pyramids?

There is ample evidence (referred to in Chapter IX.)—although, alas! the structures in Babylonia, being generally built in brick and not in stone, no longer remain, as do those erected in Egypt. Still, in spite of the absence of the possibility of a comparative study, research has shown that in the whole region to the north-east of Egypt the temenos walls of temples and the walls of towns run east and west; and though at present actual dates cannot be given, a high antiquity is suggested in the case of some of them. Further, as has been already pointed out, the temples which remain in that region where stone was procurable, as at Palmyra, Baalbek, Jerusalem, all lie east and west.

But more than this, it is well known that from the very earliest times pyramidal structures, called ziggurats, some 150 feet high, were erected in each important city. These were really observatories; they were pyramids built in steps, as is clearly shown from pictures found on contemporary tablets; and one with seven steps and of great antiquity, it is known, was restored by Nebuchadnezzar II. about 600 B.C. at Babylon.

STATUE OF CHEPHREN, FOUND IN THE TEMPLE NEAR THE SPHINX.

A careful study of the historical references to the various pyramids built in Egypt leaves it beyond doubt that the step pyramids are the oldest. They could, then, most easily have been constructed on the Babylonian model, and in this fact we have an additional argument for the intrusion of the pyramid builders into Egypt from Babylonia.

But did this equinox-worshipping, pyramid-building race live at anything like the time required?

There is no doubt now in the minds of scholars that the evidence is conclusive that among the kings of Babylonia were the following:—[1]

	B.C.
Entenna	4200
Naram-Sin	3800
Sargon I.	3750

The date of the earliest known pyramid in Egypt may perhaps be put down as about 3700 B.C. (Brugsch), or 4200 B.C. (Mariette).

Hence it seems that a third line of evidence is in favour of the Babylonian intrusion. There was undoubtedly an equinox-worshipping, pyramid-building race existing in Babylonia at the time the Egyptian pyramids are supposed to have been built.

Another connecting link is found in the statues of Chephren discovered in the temple at the pyramids, and at Tel-loh (ancient Lagash) by M. de Sarzec in 1881. This last find consisted of some large statues of diorite, and the attitude is nearly identical with that of Chephren himself as represented in the statues in the museum of Gîzeh.

This indicates equality in the arts, and the possession of similar tools, in Chaldæa and Egypt about the time in question. Further it is supposed that the diorite out of which both

[1] *See* "Guide to the British Museum," p. 71.

series of statues were fashioned came out of the same quarry in Sinai. The characters in which the inscriptions are written are in what is termed "line" Babylonian—*i.e.*, they resemble pictures more than cuneiform characters; and the standard of measurement marked upon the plan of the city, which one of the figures of Tel-loh holds upon his knees, is the same as the standard of measurement of the Egyptian pyramid builders—the cubit of 20·63, not the Assyro-Babylonian cubit of 21·6.[1]

Now, although with regard to the cult of the northern stars it was impossible to decide whether the Egyptian school of astronomers came from Babylonia or from a source common to both countries, it is clear that with regard to the equinoctial cult we are limited absolutely to Babylonia as the special source. The coincidence in time of the same kind of buildings and the same art in the two countries puts a common origin out of the question.

To sum up, then, so far as we have gone, both the north-star worship and the equinoctial worship were imported into Egypt.

[1] Sayce, Hibbert Lectures, p. 33. Flinders Petrie, *Nature*, Aug. 9, 1883, p. 341.

CHAPTER XXXV.

THE ORIGIN OF EGYPTIAN ASTRONOMY (CONTINUED)—THE THEBES SCHOOL.

THE next question which arises now that we have considered the facts relating to the astronomy of Northern Egypt is one connected with the cults which we have proved to come down the Nile. Were they indigenous or imported?

Although I have put it forward with all reserve, there is evidence which suggests that the temples so far traced sacred to the southern cult are of earlier foundation than those to the north; and they are associated with Edfû and Philæ, which are known to be of high antiquity. This is one point of difference. Another is that the almost entire absence of Set temples and east and west pyramids up the river indicates that, so far as these structures go, we lack the links which astronomically and mythologically connect the Delta with Babylonia either directly or by common origin.

From Prof. Sayce it is to be gathered that the most ancient people yet glimpsed there inhabited the region at the head of the Persian Gulf, one of the chief cities being Eridu, now represented by the mounds of Abu Shahrên on the eastern bank of the Euphrates. It was founded as a maritime city, but is now far inland, owing to the formation of the delta, the alluvium of which at the present time advances about sixty-six feet a year.[1] This alone is an argument in favour of its high antiquity.

Along with the culture of Eridu went the worship of the

[1] Sayce, *op. cit.*, p. 135.

god of Eridu, the primal god of Babylonia, Ea, Ía, or Oannês, symbolised as a goat-fish, and connected in some way with the sun when in Capricornus.

This, Jensen, by his wonderful analysis (would that I could completely follow it in its marvellous philological twistings, pages 73–81) puts beyond question; and he clinches the argument by showing that our "tropic of Capricorn" of to-day—the goat still represented on our globes of to-day with a fish's tail!—was called by the Babylonians "the path followed by Ía" or in relation to Ía.

This Ía was such a great god that to him was assigned the functions of Maker of Men; he was also a great potter and art workman (p. 293), a point I shall return to presently. He eventually formed a triad with Anu and Bīl, that is, the poles of the heavens and the equator.[1]

The God of Eridu.

Let us assume that the earliest sun-god traced at Eridu was the sun-god of those early argonauts who founded the colony.

We are told that this god was the son of Ía, and that his name was Tammuz; he was in some way associated with Asari (? Osiris) (Sayce, p. 144), who, according to Jensen, represented the Earth (p. 195); of the Moon we apparently hear nothing.

This Tammuz (Dumuzi), we find, ultimately became "the Nergal of Southern Chaldæa, the sun-god of winter and night, who rules, like Rhadamanthos, in the lower world" (Sayce, p. 245), and as lord of Hades he was made son of Mul-lil (Sayce, p. 197).

[1] One gets the idea, from reading Professor Sayce's work, that there might have been in the earliest times a north-star-worshipping race up the valley before Ía and Sun and Moon worship were established at Eridu; and that the addition of Ía to the Bīl-Anu-worship to make one triad, and the addition of Bīl to the Ía-Asari-worship to make another, were both compromises. See Sayce, pp. 326, 347, 400.

This was at first. But what do we find afterwards?

Nergal is changed into the Midsummer Sun! (Jensen, p. 484). And finally he is changed into the Spring Sun Marduk at Babylon (Sayce, p. 144)[1] where he is recognised as the son of Ía and Duazag, that is the Eastern Mountain (Jensen, p. 237).

Now, however difficult it may be to follow these changes from the religious point of view, from the astronomical side they are not only easily explained, but might have been predicted, provided one hypothesis be permitted, namely, that the colony who founded Eridu were originally inhabitants of some country where the chief agricultural operations were carried on about the time of the Autumnal Equinox in the northern hemisphere.[2]

[1] Prof. Sayce has been good enough to inform me that he is of opinion that Marduk or Merodach was originally a local god of Babylon, and that he was identified with the son of Ea when a colony came to Babylon or founded that city, bringing with it the culture and theology of the south. In this way the sun-god of Babylon became confounded with the sun-god of Eridu. I should add that Assyriologists are not all agreed about the transitions to which I have referred.

[2] I owe to the kindness of Sir Arnold Kemball, K.C.B., the perusal of a valuable report on the agriculture of British East Africa, prepared for him by Mr. W. W. A. Fitz-Gerald. He has permitted me to print the following abstract:—" The whole of the eastern coast is affected in a greater or lesser degree by the S.W. and N.E. monsoons. The following notes deal only with the extent of coast-land lying opposite and to the north of Zanzibar and Pemba islands. The agricultural seasons on the coast-lands are two in number, and correspond with the advent of the N.E. and S.W. monsoons respectively. They are distinguished locally as the 'greater rains,' or 'Masika M'Ku;' the 'lesser rains,' or 'Masika M'dogo.' The greater rains inaugurate the most important cultivating season, commencing in March with the S.W. monsoon. Some years the sowing commences as early as the 7th, but generally speaking the average period may be given as beginning from the middle of the month, and by the first week in April all sowings of Indian corn, rice and 'mfmah' (Millet or Sorghum vulgare), the chief and staple food-stuffs of the people, are generally finished, though sowings may continue till the end of April. The heaviest fall of rain occurs in April and May, and the rain continues with gradually diminishing force to September. *Harvest* takes place in July and August, and once the grain is off the field the land is immediately cleared and prepared for sowing, in anticipation of the coming of the 'lesser rains' in *October*. The season of the lesser rains is chiefly the time for the cultivation of Gingelly oil seed, beans, and such other lesser food-stuffs. The season of the 'lesser rains' is deemed more uncertain and less to be depended upon, and the rainfall is decidedly very much smaller in comparison. The 'lesser rains' practically end in November, for though the wind continues steadily from the N.N.E., the rainfall in December, January, and February is slight and uncertain, and it is during these three months, especially the two last, that the greatest heat prevails. The

This country might lie south of the equator, and indeed we find one which answers the requirements in the region of the great lakes and on the coast opposite Zanzibar.

Such an hypothesis may at first sight appear strange, but the view that Eridu was colonised from Cush has been supported by no less an authority than Lepsius.[1] The boundaries of Cush are not defined, but they may possibly include the Land of Pun-t, from which certainly part of the Egyptian culture was derived.

Among all early peoples the most important times of the year must necessarily have been those connected with seed-time and harvest in each locality. Now the spring equinox and summer solstice south of the equator are represented by the autumnal equinox and the winter solstice to the north of it. If the colonists who came to Eridu came from a region south of the equator, they would naturally have brought not only their southern stars, but their southern seasons with them; but their springtime was the northern autumn, their summer solstice the northern winter. This could have gone on for a time, and we see that their sun-god was the god of the winter solstice, Tammuz=Nergal.

But it could only have gone on for a time; the climatic facts were against such an unnatural system,[2] and the old condition could have been brought back by calling the new winter summer, or in other words making the winter-god into the summer sun-god—in short, changing Nergal into a midsummer sun-god. This it seems they did.[3]

influence of the monsoons is considerably less than on Zanzibar Island, and the difference of rainfall may be put down as about 20 to 30 inches."

[1] Introduction to "Nubische Grammatik," 1880.

[2] Just in the same way that the Equinoctial Pyramid cult gave way in Egypt, dominated by the rise of the Nile at the solstice.

[3] I shall show subsequently that a similar change seems also to have been made at Thebes. Amen-Rā, the Summer Sun-god, was a late invention.

But why the further change of Nergal to Marduk? Because the northern races were always tending southwards, being pushed from behind, while the supply of Eridu culture was not being replenished. The religion and astronomy of the north were continually being strengthened, and among this astronomy was the cult of the sun at the vernal equinox, the springtime of the northern hemisphere, sacred to Marduk. Nergal, therefore, makes another stage onward, and is changed into Marduk!

It is also interesting to find that in Ninib, another sun-god, we have almost the exact counterpart of the Egyptian Horus. He is the eastern morning sun, the son of Asari (? Osiris), and the god of agriculture.[1]

I append here the most recent translation of the hymn to the sun-god, referred to in the Introduction :—

"O Sun (god) ! on the horizon of heaven thou dawnest,
The bolt of the pure heaven thou openest,
The door of heaven thou openest.
O Sun (god) ! thou liftest up thy head to the world ;
O Sun (god) ! thou coverest the earth with the majestic brightness of heaven."

Marduk, then, the son of Ea, or Ía, was finally as definite a spring equinox sun-god as Amen-Rā in Egyptian mythology was a summer solstice sun-god.

We have, then, the undoubted facts that in Southern Babylonia, to start with, the sun-worship had to do with the winter half of the year. As the Babylonian culture advanced northward from Eridu and met the Semitic culture, the winter season was changed for the spring equinox—that is, a worship identical with that of the pyramid builders who intruded into Northern Egypt.

The Myths of Horus and Marduk.

In my references to the myth of Horus in Chapter XIV.

[1] Jensen, pp. 195—198.

I have shown that in all probability an astronomical meaning is that the rising sun puts out the northern stars. It was also indicated that the myth was one of great antiquity, as it was formulated when Draco was circumpolar; was not simple in its nature, and probably had reference to a sun-worshipping race abolishing the cult of Set representing the northern stars.

The facts brought together in subsequent chapters show that if there were not such a myth, there should have been; for the temple evidence alone showing the antithesis between Osiris-worship and the worship of Set is overwhelming.

I have also indicated that temples built to northern stars are geographically separated from those built to southern ones, and that the former have had their axes blocked to prevent the worship.

The Horus of Edfû, who is represented as leading the victorious hosts who revenge the killing of Osiris by Set, is the ally of the southern-star worshippers whom we have traced from Thebes, possibly to Central Africa (see page 350); and if we associate the myth with the records on the walls of the temple of Edfû, and agree to the possibility of that temple having been founded in 6400 B.C. (see page 311), then there must have been an invasion of the southern peoples about that date—an invasion which reached Northern Egypt, where eventually they were conquered by the Set-worshipping race, who came, as I think I have proved, from a country to the N.E. of the Delta. The question is: Did this first colony represent the original Hor-Shesu, so-called specially because perhaps as a novelty they had *added* the worship of the sun to the worship of the moon? and was the moon the first Osiris brought in by moon-worshippers with a year of 360 days?

In Accad and Sumer, where also, according to Hommel and others, the word Osiris (Asari) has been traced, the sun-god was the daughter of the moon-god. An eye forms part both of the hieroglyphic and of the cuneiform name, and the eye was one of the symbols in the name of Osiris in Egypt. Be this as it may, we have temple evidence to show that in Egypt the worship of Set was the worship of a northern race, and that it was finally abolished by a southern one.

Now in Babylonia exactly the opposite happened. The proto-Chaldæan south-star and winter-sun cult of Eridu was ultimately changed, absorbed, and buried in the Semitic cult of the northern stars Anu and Bīl and the spring sun, first Marduk and afterwards Šamaš.

Had there been then myth-makers in Babylonia, the myth would have been the converse of the Egyptian one. There were myth makers, and precisely such a myth! It is called the Myth of Marduk and Tiāmat.

The chief change had been in the sun-god. When the northern cult conquered, the exotic worship of the autumn and winter constellations was abolished, and they were pictured as destroyed under the form of Tiāmat, although the worship was once as prominent as that of Set in Egypt. We have the later developed northern spring-sun Marduk destroying the evil gods or spirits of winter; and chief among them, of course, the Goat-fish, which, from its central position, would represent the winter solstice.

The myth, then, has to do with the fact that the autumn- and winter-sun-worship of Eridu was conquered by the spring-sun-worship of the north.

If we accept this, we can compare the Egyptian and Babylonian myths from the astronomical point of view in the following manner; and a wonderful difference in the astronomical observations made, as well as in the form, though not in the basis, of astronomical mythology in Egypt and in Babylonia is before our eyes. Astronomically in both countries we are dealing with the dawn preceding sunrise on New Year's Day, and the accompanying extinction of the stars.

But which stars? In Egypt there is no question that the stars thus fading were thought of as being chiefly represented by the stars which never set—that is, the circumpolar ones, and among them the Hippopotamus chiefly. In Babylonia we have to do with the ecliptic constellations.

Now I believe that it is generally recognised that Marduk was relatively a late intruder into the Babylonian pantheon. If he were a god brought from the north by a conquering race (whether conquering by craft or *kraft* does not matter), and his worship replaced that of Ía, have we not, *mutatis mutandis*, the exact counterpart of the Egyptian myth of Horus? In the one case we have a southern star-worshipping race ousting north-star worshippers, in the other a northern equinoctial sun-worshipping race ousting the cult of the moon and solstitial sun. In the one case we have Horus, the rising sun of every day, slaying the Hippopotamus (that is, the modern Draco), the regent of night; in the other, Marduk, the spring-sun-god, slaying the animals of Tiāmat—that is apparently the origin of the Scorpion, Capricornus, and Pisces, the constellations of the winter months, which formed a belt across the sky from east to west at the vernal equinox.

The above suggested basis of the Babylonian mythology regarding the demons of Tiāmat, established when the sun was in Taurus at the spring equinox, enables us to understand clearly the much later (though similar) imagery employed when the sun at the equinox had passed from Taurus to Aries—when the Zend Avesta was written, and after the twelve zodiacal constellations had been established. We find them divided equally into the kingdoms of Ormazd and Ahriman. Here I quote Dupuis:—

> "L'agneau est aux portes de l'empire du bien et de la lumière, et la balance à celles du mal et des ténèbres; l'un est le premier des signes supérieurs, et l'autre des signes inférieurs.
>
> "Les six signes supérieurs comprennent les six mille de Dieu, et les six signes inférieurs les six mille du diable. Le bonheur de l'homme dure sous les premiers signes, et son malheur commence au septième, et dure sous les six signes affectés à Ahriman, ou au chef des ténèbres.
>
> "Sous les six signes du règne du bien et la lumière, qui sont agneau, taureau, gémeaux, cancer, lion et vierge ou épi, nous avons marqué les états variés de l'air et de la terre, qui sont le résultat de l'action du bon principe. Ainsi on lit sous l'agneau ou sous le premier mille ces mots, printemps, zephyr, verdure; sous le taureau, sève et fleur; sous les gémeaux, chaleurs et longs jours; sous le cancer, été, beaux temps; sous le lion, épis et moissons; et sous la vierge, vendanges.
>
> "En passant à la balance, on trouve les fruits; là commence le règne du mal aussitôt que l'homme vient à cueillir les pommes. La nature quitte sa parure; aussi nous avons écrit ces mots, dépouillement de la nature; sous le scorpion on lit froid; sous le sagittaire, neiges; sous le capricorne, glace et brouillard, siège des ténèbres et de longs nuits; sous le verseau, pluies et frimas; sous les poissons, vents impétueux."

Since the great pyramids were built in the time of the fourth dynasty, it is quite clear that Eridu must have been founded long before if the transitions were anything like those I have stated.

The Argument touching η Argûs.

But there is not only evidence that at Eridu the sun-worship was at first connected with the winter solstice. It is known that there was star-worship as well; and there must have been moon-worship too, judging by the moon-god of the adjacent town of Ur.

Associated with Ía was an Ía-star, which Jensen concludes may be η Argûs. This we must consider.

Jensen concludes that the Ía-star is η Argûs on the ground that many of the texts suggest a darkening of it now and again; he very properly points out that a variability in the star is the only point worth considering in this connection, and by this argument he is driven to η, which is one of the most striking variables in the heavens, outshining Canopus at its maximum. Speaking generally, everybody would agree that obscuration by clouds, etc., would not be recorded; but if the star were observed just rising above the southern horizon only, then its absence, due to such causes, would, I should fancy, be chronicled, and it must not be forgotten that this is precisely the region where the Ía star would be observed, if all of the inscriptions referred to by Jensen are to be satisfied; its place was in "*äussersten Süden*" (page 153). It was "*das Pendant des im Nordpol des Aequators sitzenden Himmels-Bi'l*" (page 148); "*Ía's 'Ort' am Himmel liegt im Süden*" (page 26).

There is another argument. Professor Sayce in his lectures reproduces (page 437) Mr. George Smith's account of the Temple of Bel derived from a Babylonian text. The temple was oriented east and west. In a description of one of the enclosures we read that on the northern side was a temple of Ía, while on the southern side there was a temple of Bīl and Anu. This not only shows that Ía was regarded as sacred to the true south, but that the temple buildings were planned like the Egyptian ones, the light either from sun or star passing over the heads of the worshippers in the courts into the temples. (Compare temple M in the temple of Amen-Rā, page 118 *ante.*)

But η Argûs never rose or set anywhere near the south.

I have ascertained that its declination was approximately 32° S. in 5000 B.C., and increased to 42° S. by about 2000 B.C. Hence between these dates at Eridu its amplitude varied between 38° and 51° S. of E. or W. Now here we are far away from the S. point, though very near the S.E. or S.W. point, to which it is stated some of the Babylonian structures had their sides oriented.

The question arises whether there was a star which answers the other conditions. *There was a series of such stars.*

It may be here mentioned generally that the precessional movement must, after certain intervals, cause this phenomenon to be repeated constantly with one star after another.

Beginning with perhaps a sufficiently remote period, we have:—

Achernar	8000 B.C.
Phact	5400 B.C.
Canopus	4700 B.C.

These stars would appear very near the south point of the horizon at Eridu at the dates stated, and describe a very small arc above it between rising and setting at certain times of the year.

Now to go a stage further in the study of the Ía—Ea or Eridu—star, it is desirable to quote the legend concerning Ía or Oannes derived from Bêrôssos through Alexander Polyhistôr.[1]

"In the first year there appeared in that part of the Erythraean sea which borders upon Babylonia a creature endowed with reason, by name Oannes, whose whole body (according to the account of Apollodôros) was that of a fish; under the fish's head he had another head, with feet also below similar to those of a man subjoined to the fish's tail. His voice, too, and language were articulate and human; and a representation of him is preserved even to this day.

"This being was accustomed to pass the day among men, but took no

[1] Sayce, p. 131.

food at that season; and he gave them an insight into letters and sciences and arts of every kind. He taught them to construct houses, *to found temples*,[1] to compile laws, and explained to them the principles of geometrical knowledge. He made them distinguish the seeds of the earth, and showed them how to collect the fruits; in short, he instructed them in everything which could tend to soften manners and humanise their lives. From that time nothing material has been added by way of improvement to his

THE TEMPLES AT PHILÆ.

instructions. Now, when the sun had set, this being Oannes used to retire again into the sea, and pass the night in the deep, for he was amphibious. After this there appeared other animals like Oannes."

It is not necessary to give the string of "other animals" enumerated by Eusebius, but one of them is important. A companion of Anôdaphas and Odakôn shows the true reading to have been Anâdakôn—that is, Anu and Dagon. This other animal, then, clearly refers to the introduction of the northern Semitic cult, and hence the suggestion is

[1] The italics are mine.—J. N. L.

strengthened that some of the earlier "other animals" who subsequently appeared, like Ía (? Oannes), may really have been new southern stars making their appearance in the manner I have shown, and perhaps varying the cult.

The whole legend is, I think, clearly one relating to men coming from the south (?) to Eridu in ships. The

THE TEMPLE AT AMADA.

boat is turned into a "fish-man," and the star to which they pointed to show whence they came is made a god.

It is evident the intrusion was from the south, because otherwise extreme south stars would not have been in question. We have, then, got so far. The worshippers of the southern star and of the winter months, including the solstice, were certainly not indigenous at Eridu. They were probably introduced from the south, and they were sea-borne.

The next question which concerns us is, was this worship in any way connected with Egypt?

One of the most definite and striking conclusions to which the study of temples has led, is that in Southern Egypt the temple worship was limited to southern stars, and, further, that there is a chain of temples, possibly dating from 6400 B.C., and oriented to Canopus. This certainly is an argument in favour of a worship similar to that traced at Eridu.

But is there any trace of Ía or of his son, the sun-god?

This god was, as we have seen, associated in some way with Asari. I am told that students will probably agree that the connection between this word and the Egyptian Osiris is absolute. Professor Sayce informs me that the cuneiform ideograms and the hieroglyphs have the same meaning, and indicate the same root-words.[1]

Ía was represented as a goat-fish, and was a potter and "maker of men." This being so, I confess the facts relating to the southern Egyptian god Chnemu strike me as very suggestive. He is represented goat-headed, and not ram-headed, as generally stated; he is not only the creator of mankind, but he is a potter, and he is actually represented at Philæ as combining these attributes in making man out of clay on a potter's wheel. Nay, according to Bunsen, he is stated to have formed on his wheel the divine limbs of Osiris, and is styled the "sculptor of all men."[2]

I give the following extracts from Lanzoni (p. 956):—

"χNUM.—χnum [Çhnemu] significa 'fabbricatore, modellatore.' . . . Questo demiurgo apparisce come una delle più antiche divinità dell' Egitto, ed aveva un culto speziale nello Nubia nell' isola di File di Beghe e di Elephantina. . . . Esso era il dio delle cataratte, identificato al dio Nun, il Padre degli dei, il principio Umido. Il grande testo geografico di Edfu parlando di Elephantina,

[1] Professor Sayce also tells me that Asari was subsequently identified by the Semitic Babylonians with Merodach.

[2] "Egypt's Place," vol. i., p. 377.

quale metropoli del primo Nomo dell' Alto Egitto, ne ricorda la divinità, come una personificazione dell' Acqua dell' inondazione."

He is also Hormaχu, the god of the universe: The father of the father of the gods: Creator of heaven, earth, water, and mountains; a local form of Osiris. His wife was the frog-goddess, Hekt (? Serk-t).

Further, he was also regarded as presiding in some special way over water,[1] and, unlike Amen-Rā, though like Ía, he has a position among the gods of the lower world.

A sun-god, with uræus and disk, he is closely associated with Amen-Rā, and if he were one of the earliest of the South Egyptian gods this could only be by Amen-Rā being an emanation from him; the temples in any case do not afford us traces of Amen-Rā before 3700 B.C., and Chnemu is recognised as one of the oldest gods in Egypt, on the same platform as Ptah in the North. If we assume a connection with Eridu, then we are driven to the conclusion that the Eridu culture came either from Egypt or from a common source.

CHNEMU.

Here for the present the question must be left. I must be content to remark that many of the facts point to a common origin south of the equator. It is clear that if Chnemu were a sun-god of the *Winter*, brought into Egypt from without, the change to Amen-Rā is precisely what would have been certain to happen, for in Egypt the Summer Solstice, over which Amen-Rā presided, was all-important.

Anthropological Evidence.

It will be seen, then, that a general survey of Egyptian

[1] Rawlinson's "Ancient Egypt," vol. i., p. 328.

history does suggest conflicts between two races, and this of course goes to strengthen the view that the temple-building phenomena suggest two different worships, depending upon race distinctions.

We have next to ask if there is any anthropological evidence at our disposal. It so happens that Virchow has directed his attention to this very point.

Premising that a strong race distinction is recognised between peoples having brachycephalic or short, and dolichocephalic or long, skulls, and that the African races belong to the latter group, I may give the following extract from his paper:—

"The craniological type in the Ancient Empire was different from that in the middle and new. The skulls from the Ancient Empire are brachycephalic, those from the new and of the present day are either dolichocephalic or mesaticephalic; the difference is therefore at least as great as that between the dolichocephalic skulls of the Frankish graves and the predominantly brachycephalic skulls of the present population of South Germany. I do not deny that we have hitherto had at our disposal only a very limited number of skulls from the Ancient Empire which have been certainly determined; that therefore the question whether the brachycephalic skull-type deduced from these was the general or a least the predominant one cannot yet be answered with certainty; but I may appeal to the fact that the sculptors of the Ancient Empire made the brachycephalic type the basis of their works of art too."

It will be seen, then, that the anthropological as well as the historical evidence runs on all fours with the results to be obtained from such a study of the old astronomy as the temples afford us.

CHAPTER XXXVI.

GENERAL CONCLUSIONS AS TO THE NORTH AND SOUTH RACES.

It is now time to summarise the evidence concerning the north and south temple builders, including those who built pyramids as well.

To do this we must deal not only with the buildings, but with the associated mythology, or, rather, with the astronomical part of the mythology, for there seems to be very little doubt that in the earliest times, before knowledge replaced or controlled imagination, everything was mythologically everything else in turn. It is for this reason that trusting to genealogies especially seems like building on sand. That Father-ship and Son-ship in the earliest days were mythologically something quite different from what the words in their strict sense imply to-day will be agreed to by everybody; and there is evidence that many of the absolute contradictions met with, and statements which it is impossible to reconcile, may all depend upon the point of view from which the mythological statements were made.

But when astronomy helps us to *the point of view*, the mythological statements, and even the genealogies, become much clearer and unmistakable, and contradiction vanishes to a great extent; and it would seem as if genealogies *en bloc* were never propounded, hence it was a commonplace either that a god should be the father of his mother, or that he should have no father.

Thus, in one sense, Rā is father of all the gods; but

in another Ptah is the creator of the egg of the sun because Capella setting heralded sunrise at a particular time of the year; and Isis is the *mother* of Horus because Phact = α Columbæ, Serk-t = α Centauri, Mut = γ Draconis, and other stars (Isis) did precisely the same; while in another connection Isis is the sister of Osiris, and therefore the *mother* of Horus. But here the relationship depends upon the association of the moon and warning star in the morning sky. I only offer these as suggestions; similar variations might be multiplied *ad nauseam*.

But while all this proves that genealogies may be manufactured without either end or utility, we gather that the association of mythological personages with definite astronomical bodies may in time be of great help in such inquiries, and ultimately enable us to raise the veil of mystery by which these old ideas have of set purpose, and partly by these means, been hidden.

There seems no doubt that we have got definite evidence that the very oldest mythological personages were closely connected either with the sun at some special time of the year, with the moon, or with the rising and setting of some star or another. Hence we ought to be able from the temple evidence to classify the northern and southern gods.

Northern Gods and Goddesses.

GOD.	GODDESS.	
Ptah = Capella, April sun (1)	Bast-Isis =	α Ursæ Majoris.
Anubis = Northern constellations.	Taurt-Isis =	α Ursæ Majoris. γ Draconis.
Min. Khem = May Sun (2).	Menat-Isis	Spica.
Autumn Sun	Serk-t-Isis	Antares.
Spring Sun	Nit-Isis	Pleiades.

Southern Gods and Goddesses.

Osiris = Moon-god.
Chnemu = Sun-god, autumnal equinox.
Khonsu = Canopus, autumnal equinox, warner.
West horizon
followed by Serk-t = α Centauri
east horizon.

Amen-Rā	A combined north and south god, established about 3700 B.C.	Teχi-Isis Amen-t-Isis Hathor-Isis	Phact (1) Sirius (2)

The establishment of Amen-Rā gives us a fair indication of the changes which must have taken place among the early civilisations when the beginning of the year was altered. There can be no doubt, I think, that Chnemu was the first Sun-god of Southern Egypt; the cryosphinxes at Thebes are alone sufficient to prove it;[1] and if so, then the southern people must have come from a region where the autumnal equinox marked the most important time of the year for their agricultural operations. And this year had eventually to give way, as we know it did, about 3700 B.C., for one beginning at the summer solstice.

In the above list I have indicated Osiris as a Moon-god. Many inscriptions might be quoted similar to the following one:—

"Salute a te, Hesiri, il signore dell' eternità. Quando tu sei in Cielo, tu apparisci come sole, et tu rinnuovi la tua forma comme Luna."[2]

It has also to be borne in mind that the complicated head-dress, including the goat's horns, is represented in connection with Thoth Chnemu and Osiris.[3]

Later he was unquestionably a sun-god, but this would be

[1] Lanzoni also states that Amen-Rā sometimes appears with the four heads of the goat, once special to Chnemu, q.v.

[2] Lanzoni, p. 692.

[3] Rawlinson, vol. i., p. 371.

certain to happen if the southern intruders worshipped the moon in the first line.

Further, if in later times he represented both sun and moon, as he certainly did, it is not probable that he did so from the beginning. All the special symbolism refers to him as a Moon-god; he is certainly a Moon-god in the myth of Isis and Osiris, for he was cut into fourteen pieces, the number of days of the waning moon.

Now, we can easily understand an evolution beginning with a Moon-god and ending with a Sun-god. But the contrary is almost unthinkable; besides, we know that in Egypt it did not happen; the solar attributes got hardened as time went on. The calendar evidence, as we have seen, in relation to the original year of 360 days is in favour of Moon-worship, and therefore of a Moon god in the earliest times.

Further, if we accept this, the myth of Horus becomes a complete historical statement, of which parts have already been shown to refer to astronomical facts past all dispute. It is well here to give Naville's remarks upon it. It will be seen that they strengthen my view.[1]

"La 363me année de son règne, le dieu part avec son fils pour l'Égypte. Voilà donc une date précise de l'un de ces rois qui, selon les traditions égyptiennes, avaient occupé le trône de l'Égypte avant les souverains indigènes. Cette année-là, Horhut chasse Typhon de l'Égypte, et s'établit en roi sur tout le pays. Cela concorderait donc avec ce que nous disent Manéthon et Eusèbe, que, dans la première dynastie des dieux, Typhon précéda immédiatement Horus. La succession se serait faite par droit de conquête.

"Horus a avec lui des compagnons qui sont nommés partout ses suivants: les Schesou Hor. M. de Rougé a déjà fait remarquer que, dans plusieurs inscriptions, ces hommes sont considérés comme les habitants primitifs de l'Égypte, les contemporains des dynasties divines. Ce sont ces Mesennou dont il a déjà été question dans la série précédente. Le rôle qu'ils jouent dans ce récit montre, plus clairement encore, que l'époque dont il s'agit est la fin des temps mythologiques auxquels Ména devait succéder. C'est une

[1] Naville, "Mythe d'Horus," p. 8.

tradition relative aux événements qui ne doivent avoir précédé que de peu les temps historiques.

"Horhut monte dans la barque de son père, qui le suit pendant toute l'expédition, et lui donne son appui et ses conseils. Les dieux poursuivent Typhon tout le long du fleuve; Horhut livre plusieurs batailles dans des lieux qui recevront des noms propres à rappeler ses exploits, et qui seront plus particulièrement voués à son culte. C'est à Edfou qu'ont lieu les premiers combats, puis dans le 16me nome de la Haute-Égypte. Le nome de Mert, celui du Fayoum et du lac Moeris, est le théâtre de plusieurs épisodes de la lutte. C'est dans la ville de Sutenchenen, appelée ici Nanrutef, un sanctuaire important d'Osiris, que s'établissent les Schesou Hor. Enfin, lorsque Set a été chassé du nome de Chent-ab, le 14me de la Basse-Égypte, le pays est délivré, et la royauté est assurée à Horhut. Son père, qui, à chaque nouvelle victoire, lui a décerné quelque honneur special, lui accorde d'être représenté sous la

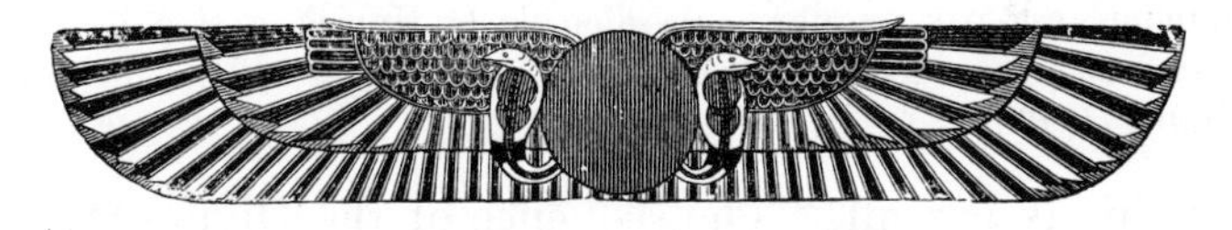

THE WINGED SOLAR DISK.

forme du disque ailé, ou du scarabée, sur tous les temples de la Haute et de la Basse-Égypte. Horus devient le seigneur des deux régions, s'assied dans un sanctuaire où il est adoré comme Horchuti, avec qui il finit par se confondre.

"Telle est cette seconde légende, bien mieux caractérisée que la première, car elle est rattachée à des localités connues et à une époque déterminée. Elle me semble même assez claire pour qu'on puisse y voir une tradition, qui aurait à sa base un fait historique. Set est un dieu bien connu dans l'histoire d'Égypte; c'est le dieu des ennemis, et particulièrement des populations sémitiques, qui conquirent une fois le pays et le mirent souvent en danger. Si nous considérons qu'il est chassé par Horus, le dieu qui lui a succédé dans la royauté, et par les habitants primitifs du pays à un moment donné des annales divines, n'est-il pas naturel d'expliquer ce mythe par une guerre entre les Égyptiens venus de Nubie, et les Sémites qui auraient été chassés du pays; soit que cette guerre soit plus ancienne que les temps historiques, soit que, venue plus tard, elle ait passé dans le domaine de l'histoire légendaire? Les textes relatifs aux dynasties divines sont encore trop rares pour que nous puissions pousser très-loin ces recherches. Le temple d'Edfou nous fournira peut-être un jour de nouvelles indications sur ces époques préhistoriques, et sur l'origine si mysterieuse de la civilisation de l'Égypte."

In another passage Naville remarks:

"Typhon n'est pas simplement le dieu du mal, l'adversaire personnel d'Osiris, c'est un souverain qui occupe avec ses alliés la plus grande partie de l'Égypte depuis Edfou jusqu'à l'Orient du Delta."[1]

It was suggested (page 154) that Horus slaying Set represented by a hippopotamus was a reference to a time antecedent to 5000 B.C., when the constellation of Draco was circumpolar; and we now learn from Chapter XXXII. that Set represented the Northern-Star worship brought in from the N.E.

Horus, then, represented a conquering force coming from the South.

He was recognised as a Southern god. Naville remarks:

"Horchuti est par excellence le dieu de la Nubie; c'est à lui que sont consacrés plusieurs des temples pharaoniques qui existent le long du Nil entre Ouadi-Halfa et Philæ."[2]

But this is not all. The sequence of the Divine Dynasties is as follows, according to Maspero:—[3]

Atmu.
Rā
Shou
Sibou [Seb]
Osiris
Set
Horus

Neglecting the first four, we find Osiris preceding Set, and are driven to the conclusion that in Osiris, in this connection, we are dealing with the Moon, for the Sun-gods Atmu and Rā head the list. Besides, the worship of Set did not kill the worship of the *Sun*, for the power of Rā finally became paramount.

We must hold, then, that the Southern Sun-god Horus, the son of Osiris, was the son of a Moon-god, and it becomes necessary to inquire if such an idea occurred to other early peoples. Professor Sayce[4] tells us—

[1] "Mythe d'Horus," p. 7.
[2] "Mythe d'Horus," p. 7.
[3] "Hist. Anc.," p. 33.
[4] "Hibbert Lectures," p. 155.

"According to the official religion of Chaldæa, the Sun-god was the offspring of the Moon-god," and he adds, "Such a belief could have arisen only where the Moon-god was the supreme object of worship. . . . . . To the Semite the Sun-god was the lord and father of the gods."[1]

If we, then, with this precedent, are prepared to take Osiris as the Moon-god of the Southern race, there is no doubt that the first Sun-god was Chnemu, and the first Southern Star-god —the star which heralded sunrise at the Autumnal Equinox— Khonsu (Canopus). Thoth also must be named, for it is certain that the Calendar which he leads was of Southern origin, because New Year's Day at the Summer Solstice was heralded first by Phact and afterwards by Sirius, both Southern stars.

There is likewise ample temple evidence to show that the Autumnal Equinoctial Sun was also heralded, and in even earlier times, first by Canopus and next by α Centauri, and it becomes a question whether the original moon-calendar of Thoth did not refer to a year beginning at the Autumnal Equinox. This is a suggestion resulting from later inquiries, and hence I have not referred to it in the chapters on the year.

And here, perhaps, in their dependence upon the Moon-god Osiris, we find the real reason that Khonsu and Thoth have lunar instead of solar emblems; Thoth led the initial lunar year, Khonsu only heralded the advent of the son of the Moon.

If this be so, before the foundation of the temple of Annu by "la grande tribu des Anou,"[2] the Southern (originally Moon-worshipping) race had already made its appearance in force in Northern Egypt, otherwise the divine dynasties would not have included Osiris; we need not be astonished that the temple evidence has disappeared there. The most northern ancient temple of Osiris was at Abydos; that also has gone, while those

[1] In modern German, even, the Moon is masculine and the Sun feminine.

[2] Maspero, *op. cit.*, p. 14.

at Philæ and Edfû remain, the latter, at some time subsequent to its original foundation, dedicated to a *female* Horus.

These things being presumed, we can now bring together in a working hypothesis the temple evidence so far as it bears upon the mythology and inter-action of the North- and South-Star worshippers.

Date B.C.

6400 A swarm from the south with Osiris, Thoth, Khonsu } Moon Gods. Chnemu (Sun God). come down the River.

They find a population worshipping Rā and Atmu. Possibly they were merely worshippers of the dawn and twilight.

The Moon worship is accepted as an addition, and *the divine dynasty of Osiris* begins.

The swarm brings a lunar year of 360 days with it, and the Egyptian Calendar beginning I. Thoth commences.

They build temples at Amada, Semneh, Philæ, Edfû, and probably Abydos. All these were probably Osiris temples, so called because Osiris, the Moon-god, was the chief deity, and they were used for the determination of the Sun's place at the Autumnal Equinox, at which time their lunar year probably began.

5400 A swarm, or swarms, from the N.E. One certainly comes by the Red Sea, and founds temples at Redisieh and Denderah; another may have come over the isthmus and founded Annu. They bring the worship of *Anu.*[1]

The *Divine dynasty* of Set is founded, and we can imagine religious strifes between the partisans of the new northern cult and the southern moon-worshippers.

These people might have come either from North Babylonia, or other swarms of the same race may have invaded North Babylonia at the same time.

±5000 [This date is fixed by Hippopotamus not being circumpolar after it. It might have been much earlier, but not much later.].

Horus with his "blacksmiths" comes down the river to revenge his "father Osiris" by killing his murderer Set (the Hippopotamus). The 6400 B.C. people, who came from the South, had been worsted by the last (5400 B.C.) swarm from the N.E., and have sent for southern assistance.

[1] *Annu* and *An* (Denderah): (? "la grande Tribu des Anou" of Maspero ?)

The South people by this time had become Sun-worshippers, and "Osiris" now means Sun as well as Moon.

The N.E. people are beaten, and there is an amalgamation of the *Original* and Southern cults. The N.E. people are reduced to second place, but Set is retained, and *Anubis* looks after sepulchres, soon to be replaced by Osiris as Southern priestcraft prevails. The priestly headquarters now are at Annu and Abydos. At the former place we have an amalgamated cult representing Sun and N. Star gods. At Abydos Osiris (changed into a Sun-God) is supreme.

Pyramid Times [Mariette 4200, Brugsch 3700.] Another swarm from N.E., certainly from Babylonia this time, and apparently by isthmus only, since no E.-W. temples are found on Red Sea roads.

They no longer bring Anu alone. There is a Spring Equinox Sun-God.

3700 Southern people at Barkal and Thebes in force; temple-building on a large scale. Chnemu begins to give place to Amen-Rā. Still more blending between *original* and Southern peoples.

3500 Final blending of North and South cults at Thebes. Temples founded there to Set and Min, on the lines of Annu and An.

3200 Establishment of worship of Amen-Rā at Thebes. Supremacy of Theban priests.

CHAPTER XXXVII.

THE EGYPTIAN AND BABYLONIAN ECLIPTIC CONSTELLATIONS.

I HAVE already, in Chapter XXXII., pointed out that at Annu we seemed limited to Set as a stellar divinity; so soon as pyramid times are reached, however, this was changed, and we found the list of the gods increased, and the worship of the sun and of stars in the constellations of the Bull and Scorpion went on, if it was not begun, in Egypt, in pyramid times. These constellations were connected with the equinoxes; and associated with the introduction of these new worships in pyramid times was the worship of the bull Apis.

The first question which now arises is, When were any ecliptic constellations established in Babylonia? and next, Which were they?

Jensen, in his "Kosmologie der Babylonier," tells us that there is some very definite information relating not only to Taurus and Scorpio, but to Capricornus and other winter constellations; and, as in Egypt so in Babylonia, for the first references to the constellations we have to refer to the religion and the mythology.

So far as I have been able to gather, any myth like the Egyptian myth of Horus, involving combats between the sun and circumpolar star gods, is entirely lacking in Babylonia, but a similar myth in relation to some of the ecliptic constellations is among the best known. Jensen shows that the first notions of the Babylonian constellations are to be got by studying the sun-gods, and especially the mythic war between the later sun-god Marduk and the monster Tiāmat.

I have already referred to Marduk; he is the Spring

Sun-God, and it has also been stated that the greatest god of ancient Babylonia, Ía of Eridu, was connected with the constellation of Capricornus.

Marduk represented the constellation of the Bull. Here I quote Jensen:—[1]

"It has already been suggested that the Bull is a symbol of the Spring-Sun *Marduk*; that he was originally complete; that he at one time extended as far as the Fish of *Ía*, *i.e.* the western Fish; that the Fish of *Ía*, out of which the sun emerged at the end of the year in ancient times to enter Taurus, is to represent *Ía*, the God of the Ocean, out of which his son *Marduk*, the early sun, rises daily; finally, that a series of constellations west of the Fish(es) is intended to represent symbolically this same ocean. *Marduk* is on the one hand, as early sun of the day (and the year), the son of *Ía*, the god of the world-water."

As to the sun-god Marduk, then, he represents the sun at the vernal equinox, when the sunrise was heralded by the stars in the Bull.

But what, then, are the fish of *Ía* and the other constellations referred to? They are all revealed to us by the myth. They are the Southern ecliptic constellations.

Tiāmat.

Tiāmat, according to Jensen, means initially the Eastern Sea (p. 307). This was expanded to mean the "Weltwasser" (p. 315), which may be taken to mean, I suppose, the origin of the Greek ὠκεανὸς, and possibly the overlying firmament of waters. These firmamental waters contain the southerly ecliptic constellations, the winter and bad-weather signs— the Scorpion, the Goat-fish, and the Fish among them.

[1] *Op. cit.*, p. 315.

It must be pointed out that these southerly constellations were associated with the God of Eridu *in his first stage.*

The Constellations referred to in the Myth of Marduk and Tiāmat.

We are indebted to the myth, then, for the knowledge that when it was invented, not only the constellations Bull and Scorpion, but also the Goat and Fishes had been established in Babylonia.

This argument is strengthened by the following considerations suggested by Jensen:—

"We look in vain among the retinue of Tiāmat for an animal corresponding to the constellations of the zodiac to the east of the vernal equinox. This cannot be accidental. If, therefore, we contended that the cosmogonic legends of the Babylonians stood in close relationship to the phenomena of sunrise on the one hand and the entrance of the sun into the vernal equinox on the other—that, in fact, the creation legends in general reflect these events—there could not be a more convincing proof of our view than the fact just mentioned. The three monsters of Tiāmat, which *Marduk* overcomes, are located in the 'water-region' of the heavens, which the Spring-Sun *Marduk* 'overcomes' before entering the (ancient) Bull. If, as cannot be doubted, the signs of the zodiac are to be regarded as symbols, and especially if a monster like the goat-fish, whose form it is difficult to recognise in the corresponding constellation, can only be regarded as a symbol, then we may assume without hesitation that at the time when the Scorpion, the Goat-Fish, and the Fish were located as signs of the zodiac in the water-region of the sky, they already played their parts as the animals of Tiāmat in the creation legends. Of course they were not taken out of a complete story and placed in the sky, but conceptions of a more general kind gave the first occasion. It does not follow that all the ancient myths now known to us must have been available, but certainly the root-stock of them, perhaps in the form of unsystematic and unconnected single stories and concepts."

There is still further evidence for the constellation of the Scorpion.

"A Scorpion-Man plays also another part in the cosmology of the Babylonians. The Scorpion-Man and his wife guard the gate leading to the Māšu mountain(s), and watch the sun at rising and setting. Their upper part reaches to the sky, and their *irtu* (breast ?) to the lower regions (Epic of Gistubar 60,9). After Gistubar has traversed the Māšu Mountain, he reaches the sea. This sea lies to the east or south-east. However obscure these conceptions may be, and however they may render a general idea impossible, one thing is clear, that the Scorpion-Men are to be imagined at the boundary between land and sea, upper and lower world, and in such a way that the upper or human portion belongs to the upper region, and the lower, the Scorpion body, to the lower. Hence the Scorpion-Man represents the boundary between light and darkness, between the firm land and the water region of the world. *Marduk*, the god of light, and vanquisher of Tiāmat, *i.e.* the ocean, has for a symbol the Bull=Taurus, into which he entered in spring. This leads almost necessarily to the supposition that both the Bull and the Scorpion were located in the heavens at a time when the sun had its vernal equinox in Taurus and its autumnal equinox in Scorpio, and that in their principal parts or most conspicuous star groups; hence probably in the vicinity of Aldebaran and Antares, or at an epoch when the principal parts of Taurus and Scorpio appeared before the sun at the equinoxes."

If my suggestion be admitted that the Babylonians dealt not with the daily fight but with the yearly fight between light and darkness—that is, the antithesis between day and night was expanded into the antithesis between the summer and the winter halves of the year—then it is clear that at the vernal equinox Scorpio setting in the west would be watching the sunrise; at the autumnal equinox rising in the east, it would be watching the sunset; one part would be visible in the sky, the other would be below the horizon in the celestial waters. If this be so, all obscurity disappears, and we have merely a very beautiful statement of a fact, from which we learn that the time to which the fact applied was about 3000 B.C., if the sun were then near the Pleiades.

Jensen, in the above-quoted passage by implication, and in a subsequent one directly, suggests that not all the zodiacal constellations were established at the same time. The Babylonians apparently began with the easier problem of having six constellations instead of twelve. For instance, we have already found that to complete the present number, between

Scorpio Capricornus Pisces

we must interpolate

Sagittarius Aquarius.

Aries and Libra seem also to be late additions according to Jensen, who writes:—

"We have already above (p. 90) attempted to explain the striking phenomenon that the Bull and Pegasus, both with half-bodies only, ἡμίτομοι, enclose the Ram between them, by the assumption that the latter was interposed later, when the sun at the time of the vernal equinox was in the hind parts of the Bull, so that this point was no longer sufficiently marked in the sky. Another matter susceptible of a like explanation may be noted in the region of the sky opposite to the Ram and the Bull. Although we cannot doubt the existence of an eastern balance, still, as already remarked (p. 68), the Greeks have often called it χηλαὶ 'claws' (of the Scorpion), and according to what has been said above (p. 312), the sign for a constellation in the neighbourhood of our Libra reads in the Arsacid inscription 'claw(s)' of the Scorpion. These facts are very simply explained on the supposition that the Scorpion originally extended into the region of the Balance, and that originally α and β Libræ represented the 'horns' of the Scorpion, but later on, when the autumnal equinox coincided with them, the term Balance was applied to them. Although this was used as an additional name, it was only natural that the old term should still be used as an equivalent. But it also indicates the great age of a portion of the zodiac."

Let us suppose that what happened in the case of Aries and Libra happened with six constellations out of the twelve: *in other words, that the original zodiac consisted only of six constellations.*

Taurus		Crab (or Tortoise)		Virgin (or ear of corn)		Scorpion		Capricornus		Pisces	
	Gemini		*Lion*		*Libra*		*Sagittarius*		*Aquarius*		*Aries*

The upper list not only classifies in an unbroken manner

the Fish-Man, the Goat-Fish, the Scorpion-Man, and Marduk of the Babylonians, but we pick up all or nearly all of the ecliptic stars or constellations met with in early Egyptian mythology, Apis, The Tortoise,[1] Min, Serk-t, Chnemu, as represented by appropriate symbols.

Further, the remarkable suppression or small representation of the Lion in both the more ancient Babylonian and Egyptian mythology is explained. I have shown before how the Babylonians with an equinoctial year would take slight account of the solstice, while it also follows that the Egyptians, who were wise enough not to use zodiacal stars for their warnings of sunrise, for the reason that stars in the brighter light of dawn near the sun are more difficult to see, might easily neglect the constellation of the Lion, as first Phact and then Sirius, both southern stars, marked for them the advent of the summer solstice; on different grounds, then, the Lion might well have been at first omitted in both countries.

Since there is a doubt as to the existence of the Lion among the first Babylonian constellations, the argument in the following paragraph would appear to refer to observations made at a later time, when totemism was less prevalent:—

"The Lion in the heavens must represent the heat of the summer. He does this most effectually when the summer solstice coincides with the constellation—that is, when its principal stars appear before the sun at the summer solstice. This happened at the time when the vernal equinox lay in Taurus, and when the principal star-group of the Bull

[1] I think I am right about the Tortoise, for I find the following passage in Jensen, p. 65, where he notes the absence of the Crab:—"Ganz absehend davon, ob dasselbe für unsere Frage von Wichtigkeit werden wird oder nicht, muss ich daran erinnern, das unter den Emblemen, welche die sogenannten 'Deeds of Salè' häufig begleiten, verschiedene Male wie der Scorpion so die Schildkröte abgebildet gefunden wird."

appeared before the sun at the time of the vernal equinox. The Water-jug (Amphora), Aquarius, must represent symbolically the watery season of winter. It does this most effectually when the winter solstice coincides with it, or its principal star-group appears before the sun at the winter solstice. This happened about the time when the vernal equinox lay in Taurus, and its principal star-group rose before the sun at the time of the vernal equinox."

Thanks to Jensen's researches, then, we have the important conclusion before us that the Babylonians, as well as the Egyptians, in early times symbolised the following constellations:—

Taurus	...	Bull.
Cancer	...	Tortoise.
Virgo	...	Ear of corn or other product representing fertility.
Scorpio	...	Scorpion.
Capricornus	...	Goat-man or goat-fish.
Pisces	...	Fish-man.

But what time was this?

We have seen that in Egypt the Bull constellation had been established possibly in the time of Mena, and that certainly both the Bull and the Scorpion had been established in pyramid times.

I have also given evidence to show that the E. and W. pyramid worship was brought from Babylonia. Now, about this date we know that Sargon I. was king of that country, and reigned at Accad or Agade, lat. 33° N., on the right bank of the Euphrates, Sippara being across the river. Here it may be mentioned that the latitudes of Eridu and Babylon are 31° N. and 32½° N. respectively, so that Agade was to the north of both.

Although the worship of Marduk—that is, the vernal

equinox Sun-god—in Babylon was much intensified when Khammurabi reigned about 2200 B.C., it is known that it existed long before; how long I cannot find. It is also very remarkable that the deities of Eridu, whenever that city was pre-eminent, were guarded by sacred bulls. We must leave it undetermined, therefore, at what date the Bull sun-god was established; but it seems certain, on the above grounds, that it must have been before pyramid times.

But we are not limited to the above line of evidence. There are astronomical considerations which will help us. For the purpose of noting the validity of the argument based upon them, a slight reference is necessary to the change of the equinoctial point along the ecliptic.

By the precessional movement, the position of the sun in the ecliptic at an equinox or solstice sweeps round the ecliptic in about 25,000 years. Now if we suppose twelve ecliptic constellations of equal size—that is, 30° long ($30° \times 12 = 360°$)—the time it would take the sun's place at the vernal equinox to pass through one constellation would be ($\frac{25,000}{12} =$) 2083 years. If the constellation of the Bull were twice as long formerly as it is now (when the constellations are twice as numerous), of course this period would be doubled.

So that the statement that the sun at the equinox was in the Bull does not help us very much to an actual date, and the constellation of the Lion could have been established 2000 years after the Bull, and yet have marked the summer solstice.

Further, if all the stars of the Bull (speaking generally) are seen at dawn—that is, before the sun rises—the sun has not yet reached the Bull. We can then, at all events, fix a minimum of time. The sun's longitude at the vernal equinox being always 0, the longitude of the most easterly part of the constellation, assuming this part not to have been changed, will give us the number of years that have elapsed.

I now go on to state Jensen's view as to the date of the introduction of the god Marduk into Babylonian mythology, or, in other words, of the worship of the spring-tide sun.

Jensen remarks:—

"It may safely be assumed that the constellations of the Scorpion and the Bull actually originated at the latest at a time when the autumnal and vernal equinoctial points respectively coincided with their principal stars. But this was the case more than 4900 years ago. But if we assume that Taurus and Scorpio were given their names at a time when their main stars rose before the sun at the time of the vernal and autumnal equinoxes respectively, we should obtain as the date of the establishment of the constellations of Taurus and Scorpio in the skies about the year –5000.[1] According to Dr. Tetens, the sun stood at the tips of the horns of the Bull at the commencement of spring 6000 years ago. At this time, therefore, Taurus had completely risen above the eastern horizon at sunrise.

"Since it is not inconceivable that in the delineation of the first signs of the zodiac a name was attached to a constellation of the ecliptic emerging from behind the sun, and apparently more or less connected, the name being such as to indicate symbolically the beginning of the spring then occurring, the time, about 1400 B.C., might also be that of the introduction of the Bull (and the Scorpion). But it is, of course, not necessary that this should have occurred at one of the three epochs mentioned; this is, indeed, highly improbable, and the process must be regarded as follows:

[1] According to a communication of Dr. Tetens, Aldebaran rose heliacally at the beginning of spring for Babylon 6900 years ago.

When the idea was conceived of indicating symbolically the beginning of spring in the sky—whether the idea originated in the brains of the masses or in that of a learned scholar, whether it had a mythological or a more scientific basis—a name was given in the first instance to the region in which the sun was at the beginning of spring, or to that west of it, the name denoting symbolically the beginning of spring. This, of course, does not exclude the possibility that more eastward portions of the ecliptic, whose stars were less prominent, were included in this name. From this we may conclude that Taurus did not originate later than –3000, for at that time Aldebaran, its principal star, stood east of the sun at the beginning of spring. Hence it would follow that our creation legends are, at least in part, just as old."[1]

It may, then, be gathered from the above that the constellations of the Bull and the Scorpion were recognised as such at the same early date both in Babylonia and Egypt; and to these we may add the Tortoise (our present Cancer) and some of the southern constellations. Further, that the date of their establishment was certainly not later than, say, 4000 B.C., and probably much earlier.

With regard to the complete ecliptic, the information seems meagre both from Babylonia and from Egypt in early times. I have already referred to the Egyptian decans, that is, the lists of stars rising at intervals of ten days. The lists will be found in Lepsius and in Brugsch's "Astronomische und

[1] With regard to these legends Jensen writes: "Now it is remarkable that the oldest historical king about whom the Babylonians know anything, Sargon of Agadi (?) is said to have lived about 3750 B.C.—*i.e.*, 5639 years ago—and that his son is called Narām-Sīn= 'favourite of Sīn,' the moon-god. And if we bear in mind that the zodiac with its signs plays into the Babylonian legends of creation, and that the Hebrew cosmogonic legends are derived from these, it is for us even more remarkable that the Jews place the creation of the world 5649 years ago, however much the figures derived from the Bible, according to other computations and traditions, may depart therefrom. Whether this is accidental or not, I do not profess to judge."

Astrologische Inschriften," but the stars have not been made out. In later times in Babylonia—say 1000 B.C.—the following list represents the results of Jensen's investigations:—

(1) Perhaps Aries (="leading sheep").
(2) A "Bull (of the Heavens)"=Aldebaran or (and) = our Taurus.
(3) Gemini.
(4) ?
(5) Perhaps Leo.
(6) The constellation of the "Corn in Ears"=the Ear of Corn. [Spica.]
(7) Probably Libra, whose stars are, however, at least in general, called "The Claw(s)" (*i.e.*, of the Scorpion).
(8) The Scorpion.
(9) Perhaps Sagittarius.
(10) The "Goat-fish"=Caper.
(11) ?
(12) The "Fish" with the "Fish band."

A few hundred years later, we learn from the works of Strassmeyer and Epping, a complete chain of twenty-eight stars along the ecliptic had been established, and most careful observations made of the paths of the moon and planets, and of all attendant phenomena. The ecliptic stars then used in Babylonia were as follows:—[1]

1.	η Piscium.	15.	α Leonis.
2.	β Arietis.	16.	ρ Leonis.
3.	α Arietis.	17.	β Leonis.
4.	η Tauri.	18.	β Virginis.
5.	α Tauri.	19.	γ Virginis.
6.	β Tauri.	20.	α Virginis.
7.	ζ Tauri.	21.	α Libræ.
8.	η Geminorum.	22.	β Libræ
9.	μ Geminorum.	23.	δ Scorpionis.
10.	γ Geminorum.	24.	α Scorpionis.
11.	α Geminorum.	25.	δ Ophiuchi.
12.	β Geminorum.	26.	α Capricorni.
13.	δ Cancri.	27.	γ Capricorni.
14.	ε Leonis.	28.	η Capricorni.

[1] "Astronomisches aus Babylon," pp. 117-133.

In Egypt, dating from the twentieth dynasty (1100 B.C.), is a series of star tables which have puzzled Egyptologists from Champollion and Biot downwards. These observations are recorded in several manuscripts found in tombs; they seem to have been given as a sort of charm to the people who were buried, in order to enable them to get through the difficulties of the way in the nether world.

The hieroglyphs state that a particular star of a particular Egyptian constellation is seen at a particular hour of the night. We have twelve lines representing the twelve hours of the night, and it is stated that we have in these vertical lines the equivalent of the lines in our transit instruments, and that the reference "in the middle," "over the right eye," "over the right shoulder," or "over the left ear," as the case may be, is simply a reference to the position of the star.

Were this confirmed, one of the remarkable things about the inquiry would be that the Egyptians did not hesitate in those days to make a constellation cover very nearly 90° of right ascension, showing that they wished to have as few constellations, including as many stars, as possible. But the best authorities all agree that these are tables of stars rising at different hours of the night, and a small constellation near the pole might have taken many hours to rise.

The observations were made on the 1st and 16th of every month. The chief stars seem to be twenty-four in number, and it looked at first as if we had really here a list of priceless value of twenty-four either ecliptic or equatorial stars, similar to the decans to which reference has already been made.

Unfortunately, however, the list has resisted all efforts to completely understand it. Whether it is a list of risings or meridian passages even is still in dispute. Quite recently, indeed, one of the investigators, Herr Gustav Bilfinger,[1] has not hesitated to consider it not a list of observations at all, but a compilation for a special purpose.

"The star-table is intended to carry the principle of time into the rigid world of the grave, and represents over the sepulchral vault 'the eternal horizon,' as the ancient Egyptians so aptly styled the grave, an imitation of the sky, a compensation for the sky of the upper world with its time-measuring motion; yet the idea here is bolder, the execution is more artificial and complicated, since the sculptor endeavoured to combine the daily and the annual motion of the celestial vault in *one* picture; wanted to transfer into the grave the temporal frames in which all human life is enacted. This endeavour to represent by one configuration both motions and both chronological units explains all the peculiarities and imperfections of our star-table.

"The simplest means of representing both motions was found in the stars, which circle the earth in the course of a day, and indicate the year by the successive appearance of new stars in the morning twilight. If the same stars were to serve both purposes in one representation, it was necessary to take twenty-four stars which rose at intervals of fifteen days, since only such followed each other at an average distance of 15°, and were therefore useful for showing the hours.

"If the calendar-maker really possessed a list of the twenty-four principal (zodiacal) stars, the course of the year was indicated thereby; but since he also wanted to represent the daily motion, he might with some justice have composed each night out of eleven of these stars, since the stars' risings are only visible during the ten middle hours of the night.

1 "Die Sterntafeln in den ägyptischen Konigsgräbern von Bibân el-Molûk," von Gustav Bilfinger (p. 69).

But ten hours would not have adequately represented the night, since this was thought of as a twelve hours' interval.

"There was a way out of it—viz., to call hora 0 'sunset,' hora 12 'sunrise,' which would have been a simple and correct solution if the division of the night into twelve parts for practical purposes had been aimed at. But this expedient he could not adopt, because he could or would only operate with stars, and the notions of sunrise and sunset found no place in his tables. Thus he was forced to *falsify* the customary division of the hours, by squeezing the twelve hours of the night into the time during which star risings are visible—viz., the dark night exclusive of twilight. On the other hand, he could not, with his principal stars at intervals of 15°, divide his night, shortened as it was by two hours, into twelve parts, and thus he was obliged to make use of two or three auxiliary stars, as we have proved in detail above, and thus yet more to disfigure the hour-division, since thereby the lengths of the hours were made very variable. These are then two things which we must not regard as peculiarities of ancient Egyptian reckoning, but as a consequence of the leading idea of our table, which did not intend to facilitate the division of the night into twelve parts by star observations, but was calculated, by the connection of thirteen stars with thirteen successive moments, to create the idea of the circling host of stars and thence the course of the night."

I give an abstract of the list of the twenty-four principal stars and the Egyptian constellations in which they occur :—

1. Sahu=Orion.
2. Sothis=Sirius.
3. The two stars.
4. The stars of the water.
5. The lion.
6. The many stars.

7. Mena's herald.
8. Mena.
9. Mena's followers.
10.–14. Hippopotamus.
15.–20. Necht.
21. Ari.
22.–23. Goose.
24. Sahu = Head of Orion.

It will be seen that even this Egyptian star-list is very indeterminate. It is known that Sahu is the name for the constellation of Orion. The hippopotamus represents Draco, and probably Necht another northern constellation. There are indications, too, that Mena may symbolise Spica, with which star we have seen Min-worship associated. Further than this the authorities do not venture at present to go.

CHAPTER XXXVIII.

THE INFLUENCE OF EGYPT UPON TEMPLE-ORIENTATION IN GREECE.

In the final pages of this book I have to show that recent investigations have put beyond all doubt the fact that the astronomical observations and temple-worship of the Egyptians formed the basis first of Greek and later of Latin temple-building.

I have indicated in a former chapter that in our own days, and in our own land, the idea of orientation which I have endeavoured to work out for Egypt still holds its own. It was more than probable, therefore, that we should find the intermediate stages in those countries whither by universal consent Egyptian ideas percolated. Among these, Greece holds the first place, as it was the nearest point of Europe to the Nile Valley.

Before we study the orientation of the Greek temples, let us endeavour to realise the conditions of those Greek colonists who, filled with the Egyptian learning; impressed with the massive and glorious temples in which they had worshipped; favoured, perchance, moreover, with glimpses of the esoteric ideas of the priesthood; and finally, fired with Greek ideals of the beautiful, determined that their new land should not remain altarless.

What would they do? They would naturally adapt the Egyptian temple to the new surroundings, climatic among others. The open courts and flat roofs of Egyptian temples would give way to covered courts and sloping roofs to deal with a more copious rainfall; and it is curious to note that the chief architectural differences have this simple origin.

The small financial resources of a colony would be reason good enough for a cella not far from the entrance, with courts surrounding it under the now necessary roof. The instinctive love of beauty would do the rest, and make it a *sine quâ non*

A GREEK TEMPLE RESTORED—THE TEMPLE OF POSEIDON AT PÆSTUM.

that the rosy-fingered dawn should be observable, and that the coloured light of the rising sun in the more boreal clime should render glorious a stately statue of the divinity.

It is well to take this opportunity of emphasising the transition from the Egyptian form of temple to the Greek one, in order to show how completely among many apparent changes the astronomical conditions were retained. The entrance door and the cella are always in the axis of the temple; the number of columns in the front is always even; *the door is never blocked.*

I have already pointed out that in both groups of Egyptian

THE TEMPLE OF THESEUS AT ATHENS: THE ACROPOLIS, WITH THE PARTHENON, IN THE BACKGROUND.

temples, whether furnished with a pylon or not, one goes from the entrance to the other end, which held the sanctuary, through various halls of different styles of architecture and different stages of magnificence. But in the Greek temple this is entirely changed; the approach to the temple was outside—witness the glorious propylæum of the Parthenon at Athens—the temple representing, so to speak, only the core, the Holy of Holies, of the Egyptian temple; and any magnificent approach to it which could be given was given from the outside. Be it further remarked that the propylæum was never in the fair-way of the light entering the temple.

THE EAST FRONT OF THE PARTHENON, FACING THE RISING OF THE PLEIADES.

The massive pylons of some of the Egyptian temples were useful for shading the roofless outer courts. In Greece these were no longer useful.

The east front of the Parthenon very much more resembles the temple of Denderah than it does the early Egyptian temple—that is to say, the eastern front is open; it is not closed by pylons.

The view as to the possibility of temple-orientation being dominated by astronomical ideas first struck me at Athens and Eleusis, and when I found that the same idea had been held by Nissen, and that the validity of it seemed to be beyond all question, I consulted my friend Mr. F. C. Penrose specially with regard to Greece, as I knew he had made a special study of some of the temples, and that, he being an astronomer as well as an archæologist (for, alas! they are not, as I think they should be, convertible terms), it was possible that his observations with regard to them included the requisite data.

I was fortunate enough to find that he had already determined the orientation of the Parthenon with sufficient accuracy to enable him to agree in my conclusion that that temple had been directed to the rising of the Pleiades. He has subsequently taken up the whole subject with regard to Greece in a most admirable and complete way,[1] and has communicated papers to the Society of Antiquaries (February 18, 1892), and more recently to the Royal Society (April 27, 1893) on his results.[2]

These results are so numerous and complete that it is now quite possible to trace the transition from Egyptian to Greek temple-worship, and this, with Mr. Penrose's full permission, I propose to do in this chapter.

But, in the first instance, I am anxious to state that Mr. Penrose was soon convinced that in Greece, as in Egypt, the stars were used for heralding sunrise. He writes:—

[1] In the lists of temples which follow, all the orientations were obtained from azimuths taken with a theodolite, either from the sun or from the planet Venus. In almost every case two or more sights were observed, and occasionally also the performance of the instrument was tested by stars at night. The heights subtended by the visible horizon opposite to the axes of the temples were also observed.

[2] *See Nature*, February 25, 1892, and May 11, 1893.

"The object the ancients had in using the stars was to employ their rising and setting as a clock to give warning of the sunrise, so that on the special feast days the priests should have timely notice for preparing the sacrifice or ceremonial, whatever it may have been:

"'Spectans orientia solis
Lumina *rite* cavis undam de flumine palmis
Sustulit,' etc."

I may further give an extract from a letter received from him in which he deals with the demonstration of the orientation hypothesis furnished by the Greek temples alone.

"In my paper sent to the Royal Society there was a passage which seems to make it practically *certain* that heliacal stars were connected with the intra-solstitial temples as derived from Greek sources alone, independent of the powerful aid of the Egyptian cases.

"'That the first beam of sunrise should fall upon the statue centrally placed in the adytum of a temple or on the incense altar in front of it on a particular day, it would be requisite that the orientation of the temple should coincide with the amplitude of the sun as it rose above the visible horizon, be it mountain or plain.

"'That a star should act as time-warner it was necessary that it should have so nearly the same amplitude as the sun that it could be seen from the adytum through the eastern door, if it was to give warning at its rising, or to have a similar but reversed amplitude towards the west, if its heliacal setting was to be observed; and it follows that in the choice of the festival day and the corresponding orientation, on these principles, both the amplitude of the sun at its rising and that of the star eastwards or westwards, as the case might be, would have to be considered in connection with one another.

"'From what has been said it is obvious that in the intra-solstitial temples the list of available bright stars and constellations is in the first instance limited to those which lie within a few degrees of the ecliptic, and it will be found that in the list above given and those which follow, if we omit Eleusis, where the conditions were exceptional, all but one of the stars are found in the zodiacal constellations. A very great limit is imposed, in the second place, by one of the conditions being the heliacal rising or setting of those stars from which the selection has to be made. So that, when both these combined limitations are taken into account, it becomes improbable to the greatest degree that in every instance of intra-solstitial temples of early foundation of which I have accurate particulars, being twenty-eight in number and varying in their orientation from 21° N. to 18° 25′ S. of the true east, there should be found a bright heliacal star or constellation in the right position at dates not in themselves improbable unless the temples had been so oriented as to secure this combination.'

"I have just been looking into the number of possible stars which could have been used, *i.e.* within the limits of the greatest distance from the ecliptic that could have been utilised.

"The stars which could have been utilised in addition to the seven which serve for nearly thirty temples are ten only, viz. :—

Aldebaran.	β Libræ.
Pollux.	α Libræ.
β Arietis.	α Leonis.
β Tauri.	γ Leonis.
α and β Capricorni as a group.	β Leonis.

"If the orientations had been placed at random, would not our thirty temples have made many misses in aiming at these seventeen stars, it being necessary also to hit exactly the heliacal margin? And would they have secured anything like a due archæological sequence?

"Another point is this :—

"Whenever a star less than first magnitude is used (Pleiades only excepted) it has been necessary, to secure coincidence, to give it several more degrees of sun depression than in the cases of Spica and Antares."

The problem in Greece was slightly different from that in Egypt. We had not such a great antiquity almost without records to deal with, and moreover the feast-calendars of the various temples presented less difficulty. There was no vague year to contend with, and in some cases the actual dates of building were known within a very few years.

In Greece, not dominated by the rise of the Nile, we should not expect the year to begin at a solstice, but rather at the vernal equinox. I have shown that even in pyramid times in Egypt the risings of the Pleiades and Antares were watched to herald the equinoctial sun; it is not surprising, therefore, to find the earliest temples in Greece to be so oriented. Mr. Penrose has found the following :—

				B.C.
η Tauri (The Pleiades)	Archaic temple of Minerva	Athens ...	R[1]	1530
	Asclepieion	Epidaurus ...	R	1275
	The Hecatompedon (site of Parthenon)	Athens ...	R	1150
	Temple of Bacchus...	Athens ...	R	1030
	Temple of Minerva...	Sunium ...	S	845

				B.C.
Antares	Heræum	Argos ...	R	1760
	Earlier Erechtheum	Athens ...	S	1070
	Temple at	Corinth ..	S	770
	Temple on the Mountain ... ... Jupiter Panhellenius	Ægina ...	S	630

Here we find the oldest temple in a spot which by common consent is the very cradle of Greek civilisation.

It has also been shown that in Khu-en-Aten's time the sun-temple at Tell el-Amarna was oriented to Spica. Spica, too, we find so used in Greece in the following temples:—

				B.C.	
Spica	The Heræum at	Olympia ...	R	1445	
	Nike Apteros ...	Athens ...	S	1130	
	Themis... ...	Rhamnus...	R	1092	
	Nemesis ...	Rhamnus...	R	747	
	Apollo	Bassæ ...	R	728	Eastern doorway.
	Diana	Ephesus ...	R	715	

When the sun at the spring equinox had left Taurus and entered Aries, owing to precession, in Egypt the equinoxes were no longer in question, since the solstitial year was thoroughly established, and consequently we find no temples to the new warning star α Arietis.

In Greece, however, where the vernal equinox had now been established as the beginning of the year, we find a

[1] R indicates a rising, and S a setting observation.

THE TEMPLE OF JUPITER OLYMPIUS BELOW THE ACROPOLIS, AT ATHENS, ORIENTED TO α ARIETIS.

different state of things. No less than seven temples oriented to α Arietis are already known:—

				B.C.	
α Arietis	Minerva	Tegea ...	R	1580	
	Jupiter Olympius ...	Athens ...	R	1202	
	Jupiter	Olympia ...	R	790	
	Temple (perhaps Juno)	Platea ...	S	650	
	Jupiter	Megalopolis	S	605	
	Temple at the Harbour	Ægina ...	S	580	
	Temple on Acropolis of	Mycenæ ...	R	540	Eastern doorway.
	The Metroum ...	Olympia ...	S	360	

The above are all intra-solstitial temples—that is, the

sunlight as well as the light of the star can enter them—and this enables us to note a certain change of thought brought about in all probability by the artistic spirit of the Greeks. The Egyptian temples were all dark, often with a statue of a god or a reptile obscure in the naos, and many were oriented so that sunlight never entered them. Mr. Penrose points out that almost all the Greek temples are oriented so that sunlight can enter them. Of such temples we have the following twenty-nine:—

7	examples	from	Athens.	1	example	from	Sunium.
3	,,	,,	Olympia.	1	,,	,,	Corinth.
2	,,	,,	Epidaurus.	1	,,	,,	Bassæ.
2	,,	,,	Rhamnus.	1	,,	,,	Ephesus.
2	,,	,,	Ægina.	1	,,	,,	Platæa.
2	,,	,,	Tegea.	1	,,	,,	Lycosura.
1	,,	,,	Nemea.	1	,,	,,	Megalopolis.
1	,,	,,	Corcyra.	2	,,	,,	Argos

Now in all these Greek temples, instead of the dark naos of the Egyptian building, we find the cella fully illumined and facing the entrance. Frequently, too, there was a chryselephantine statue to be rendered glorious by the coloured morning sunlight falling upon it, or, if any temple had the westerly aspect, by the sunset glow.

It was perhaps this, combined eventually with the much later invention of water-clocks for telling the hours of the night, which led to the non-building of temples resembling those at Thebes and Denderah facing nearly north; of these, however, there are scattered examples; one of very remarkable importance, as it is a temple oriented to γ Draconis 1130 B.C., built therefore not very long after the temple M at Karnak, and this temple is at Bœotian Thebes! A better proof of the influence exerted by the Egyptians over the temple-building in Greece could scarcely be imagined. As Mr. Penrose remarks:—

"Thebes was called the City of the Dragon, and tradition records that Cadmus introduced both Phœnician and Egyptian worship."

It would be very surprising, if we assume, as we are bound to do, that these temples to stars were built under Egyptian influence, that Sirius should not be represented among them, that being the paramount star in Egypt at a time when we should expect to find her influence most important in Greece. Still, I have shown already that, as the Greek year ignored the solstice, the use of Sirius as a warning star for all purposes of utility would not come in. Mr. Penrose finds, however, that, in spite of this, Sirius was used for temple-worship.

"Leaving the solar temples, we find that the star which was observed at the great temple of Ceres must have been Sirius, not used, however, heliacally —although this temple is not extra-solstitial—but for its own refulgence at midnight. The date so determined is quite consistent with the probable time of the foundation of the Eleusinian Mysteries, and the time of the year when at its rising it would have crossed the axis at midnight agrees exactly with that of the celebration of the Great Mysteries."

"It is reasonable to suppose that when, as in the case of Sirius at Eleusis, brilliant stars were observed at night, the effect was enhanced by the priests by means of polished surfaces."

Another question. Does the star follow the cult in Greece as it does in Egypt?

In Greece we find the following:—

"The star α Arietis is the brightest star of the *first sign of the Zodiac*, and would therefore be peculiarly appropriate to the temple of Jupiter. The heliacal rising of this star agrees both with the Olympieium at Athens and that at Olympia. There is a considerable difference in the deviation of the axes of these two temples from the true east; but this is exactly accounted for by the greater apparent altitude of Hymettus over the more distant mountain at Olympia.[1]

[1] With regard to a temple of Minerva using α Arietis at Tegea, Mr. Penrose writes:— "Minerva is allowed by the poets to have been able to use Jupiter's thunder, so this is no misappropriation of the star. Juno also seems to have claimed the use of α Arietis as at Samos, and at Girgenti it suits the orientation of the temple of Juno better than Spica. But Spica seems to have been connected with the worship of Juno and Diana in their more strictly female capacity."

"The Pleiades are common to the following temples of Minerva—viz., the Archaic temple on the Acropolis, the Hecatompedon, and Sunium. In the two former it is the rising, the latter the setting star.

"There must have been something in common between the temples at Corinth, Ægina, and Nemea. The two last, at any rate, are reputed temples of Jupiter."

The Greek side of the inquiry becomes more interesting when the connection between the orientation of the intra-solstitial temples and the local festivals is inquired into; in Egypt this is all but impossible at present.

A temple oriented to either solstice can only be associated with the longest or with the shortest day; if the temple points to the sunrise or sunset at any other period of the year, the sunlight will enter the temple twice, whether it points to the sunrise or sunset place.

Now Mr. Penrose finds that in Greece, as in Egypt, the initial orientation of each intra-solstitial temple was to a star, and this would, of course, secure observations of the star and the holding of an associated festival at the same time of the year for a long period. But when the precessional movement carried the star away, they would only have the sun to depend on, and this they might use twice a year. It is possible, as Mr. Penrose remarks, that

"there would have been no reason for preferring one of these solar coincidences to the other, and the feast could have been shifted to a different date if it had been thought more convenient."

He goes on to add:—

"It would appear that something of this sort may have taken place at Athens, for we find on the Acropolis the archaic temple, which seems to have been intended originally for a vernal festival, offering its axis to the autumnal sunrise on the very day of the great Panathenaia in August.

"The chryselephantine statue of the Parthenon, which temple followed on the same lines as the earlier Hecatompedon (originally founded to follow the rising of the Pleiades after that constellation had deserted the archaic temple alongside), was lighted up by the sunrise on the feast to the same goddess in August, the Synæcia, instead of some spring festival, for which both these temples seem at first to have been founded.

"The temple at Sunium, already quoted for its October star-heralded festival to Minerva, was oriented also axially to the sun on February 21, the feast of the Lesser Mysteries."

I have had to insist again and again that in the case of the Egyptian temples the stated date of foundation of a temple is almost always long after that in which its lines were laid down in accordance with the ritual. No wonder, then, that the same thing is noticed in Greece.

"In about two-thirds of the cases which I have investigated the dates deduced from the orientations are clearly earlier than the architectural remains now visible above the ground. This is explained by the temples having been rebuilt upon old foundations, as may be seen in several cases which have been excavated, of which the archaic temple of Minerva on the Acropolis of Athens and the temple of Jupiter of Olympius on a lower site are instances. There are temples also of the middle epoch, such as the examples at Corinth, Ægina, and the later temples at Argos and at Olympia (the Metroum at the last-named), of which the orientation dates are not inconsistent with what may be gathered from other sources."

The problem is, moreover, helped in Greece by architectural considerations, which are frequently lacking in Egypt: of two temples it can be shown, on this evidence alone, that one is older than the other. Such an appeal strengthens my suggestion that two of the temples of the Acropolis Hill were oriented to the Pleiades, by showing the older temple to point to an earlier position of the star group. To these Mr. Penrose adds another pair at Rhamnus, where he has found that there are two temples almost touching one another, both following (and with accordant dates) the shifting places of Spica, and still another pair at Tegea.

INDEX.

T